The Walnut Tree

The Walnut Tree

Tales of Growing and Uses

Charles Hulbert-Powell

Unicorn Press

First published in 2019 by
Unicorn Press
60 Bracondale
Norwich NR1 2BE

www.unicornpublishing.org
tradfordhugh@gmail.com

ISBN 978 1 911604 57 0

Designed by Nick Newton Design
Printed in Wales by Gomer Press Limited, Llandysul, Ceredigion

Frontispiece: Walnuts ripening in the Dordogne.

Contents

 **Foreword**

This book is a celebration of a most remarkable tree. It is well travelled, migrating at the instigation of man, principally for human beings to enjoy the nuts the tree produces and for the practical uses the timber from the tree can provide. As we shall see, this dual usage provides a wide range of foods and products. The British, as a nation, would not think of the walnut before considering the oak, the ash, the beech, the apple or the declining English elm. The French in certain areas might describe the walnut as common as the oak, the poplar, or the chestnut, but would certainly extol its value in a more enthusiastic manner than we in Britain do. The tree is common in parts of the Dordogne and near Grenoble.

Any tree is a special feature in our landscape in Britain. Trees are noticeable in our parks, in our cities, the garden, and the forest. They are pleasing to the eye but more than that, we cannot do without them. They take in carbon dioxide, they provide building materials, they provide food and they hide low height blots on the landscape like railways and roads. They provide shelter and hiding for wildlife and shade from the sun. They feed humans and animals by producing nuts and the timber provides materials for houses and many other structures including furniture. Yet some trees are more useful than others. Conifers are soft wood used in building. Hazel and chestnut are used for fencing in the United Kingdom. Poplar is used for matches, coffins and tea chests. At the other end of the scale, high quality furniture is made from trees like oak, mahogany, teak, ebony, elm, beech, rosewood and of course, the walnut.

I did not write this book with any thought that it would be a sylviculture textbook, a recipe book, an expert's guide book on gunstocks or furniture. It is meant to be a light read about a most versatile tree. I fell for the idea of writing a book by mistake. Having embarked on the growing of the walnut tree myself on my few acres in East Sussex, I found no book which brings the husbandry, the many uses of the timber, the uses and manufacture of the oil and all the recipes and possible uses of the nut itself together in one compendium. I have drawn information and observations from my own experience of planting the tree. Vita Sackville-West wrote her books and articles from the benefit of her own experience and I wanted to do the same. I have grown many trees during my time in Sussex, tried to give them the best husbandry and the limited benefit of my experience in many ways.

I have visited many knowledgeable and interesting people who have been helpful in providing their opinions and experiences in writing this book. I have tried to create a snap shot of the early life of a tree which will have an effective economic nut production life span of between 75 and 90 years. Thereafter, once its useful life as a food producer is over the tree has an afterlife in a form which is the choice of man. This form is the fashioning of the tree into timber and then into furniture, panelling or gunstocks.

The walnut is a tree that has evolved since ancient times, was involved in history and admired for centuries. In Mesopotamia, the old name for Iraq and thought to be the site of the hanging gardens of Babylon which existed in 2000 BC, the Chaldean people left clay tablets which have depictions of walnut orchards, proving that these people knew of the value of the walnut tree.

The tree travelled rather like a nomad. The intriguing migration of the walnut (*Juglans regia*) tree seems to have come from further east of the Iranian mountains; the supposed site of the Garden of Eden. Its route was west via Turkey, Greece, Italy and northwards to the British Isles and is a story in itself. The Romans were prepared to pay the Greeks for walnuts which were part of the trade between the two countries. The *Encyclopaedia Britannica* mentions the walnut in 1768, but it is thought the Romans brought the seed to Britain. In old English the translation of the walnut was 'a foreign nut'. In between Italy and Britain, evidence of walnuts was found near the Swiss lakes dating back from Neolithic times. The local history of its route is rather shrouded in mystery with only a few clues to lift the fog of ignorance.

The common walnut was also taken across the Atlantic to America where some were grafted to local indigenous stock. The common walnut migrated to the east into India and China, which is a passage to a different climate from its origin. From these basic pieces of information we are looking at a most versatile tree.

The black walnut, on the other hand, is a native of Eastern America, but is found in California, China, India, as far south as New Zealand, and sparsely in Britain. This is again a story of a remarkable migration. How did it find its way across the Pacific?

Walnuts are rarely grown in the United Kingdom as a cash crop today, although a few centuries ago I believe people like John Evelyn, the nurserymen of the 17th and 18th centuries and estate owners like Lord Petre may have seen a potential for money from the crop. My interest, living in the United Kingdom, was in part driven by publicity indicating global warming would change our weather and thus the crops we, in the United Kingdom, could grow in the future. This led me to believe that a cash crop from a walnut grove in the future might be possible. Could it be that the climate of the Dordogne would move north to the southern counties of the United

Kingdom? It might happen, but over a long period. It might not happen in my lifetime, but possibly during the lifetime of my children. It was a risk for me, because at that moment, Britain did not have the ideal climate for walnut growing or the ideal soil conditions for extensive nut production.

The reality is that the modern British walnut tree planter has amenity in mind. That means planting for show and to a lesser extent home grown food. The walnut tree can be planted in any reasonable sized British garden, but they do grow quite big and more out than up. They need space as a priority, although there are fine collections at Kew and Westonbirt where they have been planted in groups amongst other trees. Outside a national collection, I do not believe a walnut tree would be suitable for a garden of less than half an acre. Philip Waites, the head gardener at Wimpole Hall, the site of a National Walnut collection, says he likes to plant walnuts in batches of three, thinning at a later date to allow space. It is also thought that the harvest of nuts is greater, if there is more than one tree. Pollination should be more effective. Yet there are many small gardens where owners tell me that they have more than adequate harvests from just one tree.

As we shall see, planting trees is fairly easy, but after that there are excitements, frustrations and pleasures in the enterprise of looking after the tree and harvesting its nuts. In Britain we have a real variation in weather. We can experience quite violent winds and frost in the first part of the year when pollination takes place, followed by poor weather, wind and rain in September and October at the time of harvest. We have different amounts of rain in different parts of the country and occasional snow in the course of any year. There can be poor summers like 2012. That year we saw young nuts in the early spring, but after excessive wind and rain in June and July, described as the worst on record, there were very few nuts in August. That was the bad news for the year, but it must be admitted that the wet weather did encourage good growth on the trees during this season. Such growth could well benefit the trees in other years when weather conditions might be nearer normal.

In 2014 we had a very long wet and cold spring. The summer was warm and there was a prolonged period of warm weather in September. The trees had grown well and we had the first real crop from our imported trees. To us it was a triumph, but we felt we had been lucky after the cold and damp spring.

There had been reports that the climate was changing. The springs are getting earlier, we are told by climate experts. The buds of the walnut are breaking earlier as the leaf is formed. At the end of the year as autumn approaches, the leaf is changing colour and falling earlier. But we have not seen this in the French trees we have at the farm. We have observed that the leaf comes late in April and in 2014 and 2015, the leaves were still green in October behind our native oak and ash trees whose leaves had already turned brown.

High winds could also be a sign of global warming. In exposed places in August and September, a big wind can dislodge the nuts before they are ripe. The nut is formed inside an outer green skin and, unless the nut is really ripe it takes time and is difficult to remove. Inside the green skin is an exceptionally wet place which sticks to the hard surface of the nut.

There are other frustrating ways in which the crop can be reduced, if not controlled. There are evil predators such as pigeons, crows and squirrels which not only damage the tree but take most of the nuts in order to satisfy their hunger. The faint-hearted may think that a walnut project is not worth the effort, but when the crop comes, it is all worthwhile.

Three-year-old trees, well manured to improve soil condition.

Five in the evening

Sept. 19, 1708

Dear Prue,

I send you seven penny-worth of walnuts at five a penny, which is the greatest proof I can give you, at present, of my being with my whole heart,

Yours

Rich. Steele.

p.s. There are but twenty-nine walnuts.

[Richard Steele (1672–1729) essayist, who married Mary Scurlock, 'Prue', in 1707]

1 Tree planting

For most people, it is probably true that walnuts are the nuts which are hard to crack at Christmas, a treat on top of a cake, a walnut whip or an ingredient in ice cream. For many years, I had no idea, nor did I care, where they came from, their history or what you could do with them. I had seen walnut trees in gardens but not in groves, although they do exist in the UK. I had read very little about them and had certainly been told nothing about nut trees at the Royal Agricultural College where I had done a farming course.

Then one day, a friend told me that he was going to France to collect a quantity of grafted walnut trees and would I like to join in a sort of co-operative to produce nuts? The idea appealed because we had few realistic projects with any prospect of economic success on our 70 acres of poor, clay, grade three East Sussex agricultural land. I had listened to dire warnings in the early years of the 21st century about climate change from the world's climate experts. The opinion these scientists proposed was that this change threatened to turn the South of England into a dryer climate similar to that common in the south of France during the last few centuries. Yet in England I have heard the comment that our walnuts never grow very large. The reason for this could be because of soil conditions, varieties that grow here and the weather. There are indeed areas of the country where walnuts are not frequently found. One notable garden and tree expert in Cornwall told me there are few examples in the county because the climate is too wet. It is true that a wet climate is not ideal for this tree.

American growers mulch in the autumn with horse manure to help growth, a little lime could also help in very acidic areas and I believe that the spreading of wood ash around the base of the tree will help improve the soil. Potassium from the ash is a trace element known to encourage fruit and nuts.

Armed with these opinions and observations, it seemed that the planting of a small walnut grove was a challenge worth trying. The French nurserymen's selling point, was that the trees could produce nuts in five years and when in full production, should produce 50 kilograms from each tree. The prediction was right. After five years, we began to see a crop, by which I mean nuts on some of the trees in numbers. I would add that this quantity was always subject to the crows, pigeons, magpies and the dreaded tree rat (squirrel) not stealing the lot. That sounds dramatic, but sadly, it is true that if these thieves are not controlled, they can leave only a small harvest. In our first cropping year on the farm, these creatures were certainly present and were working to deplete the crop. They cannot be entirely stopped no matter how much they are controlled.

Farming is reliant on subsidies and I was pleased to hear that the new plantation would not affect the all-important single payment scheme granted by Natural England. The official who visited us thought that sheep could be allowed to graze under the trees and therefore the status of pasture could be preserved. The added benefit was that after a minimum of 30 years, the trees could be felled and the timber was likely to be valuable. The subsidy was therefore not a problem.

My friend delivered a quantity of French trees which arrived on a snowy day in February. We stored them, wetting the roots and binding them under tarpaulins, protecting them against the cold in an oast house. This oast house has its own climate and remains warm in winter which was good for the trees which were vulnerable out of the ground. When the weather improved, we planted them in a field near to the farm buildings where we could see and enjoy them and keep an eye on their progress.

We bought two types of walnuts. They were Franquetta and Fernor which had been grafted onto an unknown root stock. Franquetta has an elongated nut which is hard and wrinkled. The kernel represents 41% of the weight of the nut. Fernor is a hedgerow tree. Both varieties appeared as a 1.8 metre bare pole about five centimetres across with a substantial root bowl.

If the project means there are 50 plus trees to plant, it is a real effort to plant with a single spade. A hole 60 centimetres deep and 60 centimetres square is required. Horse manure was spread at the bottom, a little bit of wood ash was included and a small amount of fish, blood and bone meal. A layer of earth went on next and the tree was then planted. We got the holes dug in the first instance by a tractor-mounted auger at four metres apart for the hedgerow plants and seven metres for the ordinary trees.

We tied the whips to stakes which were three centimetres square and a metre long. The site we had chosen was relatively sheltered and it is possible that we might have got away without stakes. One of the problems we found was that the plastic ties tended to drop to the bottom of the tree quite soon after planting. It could have been our fault in that we fitted them poorly. Added to this, it did not take much to snap the posts, which were softwood. I learnt a lesson that if planting is done in an exposed position, a robust stake and a nail through the tie to keep the tree in position is a better way to support the tree. It should also be borne in mind that knocking a stake in after the tree is planted can damage the roots.

Beth Chatto said on a television programme celebrating many years of garden design how important it is to replicate the conditions as near as possible from where the original plant came from. The common walnut came from Georgia and the east of Turkey. There is calcareous soil in that region. Calcareous soil is found in parts of the United Kingdom, but not in our part of East Sussex. Our soil is very acidic and rhododendrons, azaleas, camellias and magnolias thrive. The chalk downs about 30 kilometres to the south and many other parts of England are alkaline and calcareous. The roots of trees will try to find nutrients as they grow into the soil, and will grow around rocks and other obstacles if there is a problem in this quest. Sometimes the tree cannot find what it needs and does not thrive. On our farm there is grey clay which is not fertile except for trees such as oak or hazel. We had to think very carefully about whether this type of soil was present on the site we planned to plant.

The pH of the soil, the measurement of acidity or alkalinity should, ideally for the walnut, be a minimum of 5 and ideally 5.5. It helps the growth of the tree to keep the pH at the right level in acidic areas. This can be done by using a slow release lime from garden centres or agricultural merchants. An alternative is to use calcified seaweed. This substance comes in bags and is spread over the root area away from the trunk. In the walnut's original natural habitat in Iran and Georgia, this chemical help is not available or necessary as the soil is ideal. In Britain we have different soils all over the country and hence the need to test the soil.

When we planted the French walnuts, we did not know how they would react to a very cold winter. It is true that walnuts can survive in hostile cold conditions but we had a bad and cold winter which lasted well into May in 2013. The spring came and the walnuts sprang to life, but some faster than others. The two trees we had bought in England a few years before the French stock was purchased produced catkins and buds much earlier. We did not see leaves until well into June.

We had not seen many deer all the time we had lived on the farm. My family have been at the farm for 40 years or more. I believe that warm winters in recent years going back to 1990 have been ideal for good reproduction of deer. They enjoyed plentiful food and places of concealment and the result has been an enormous increase in their numbers. The type of deer we have spotted in the East Sussex area are Fallow and Roe. When we planted the trees, we knew there were a number of deer in the area and that they could damage the young trees.

There was evidence of their arrival. From time to time we had reports of sightings and occasionally saw a herd of seven or eight, including an albino stag, grazing within easy range of the precious walnut grove.

In the second year of my walnut project, I found a source from which to purchase some American walnuts. They did not have the charming French names of Franquetta or Fernor. They were referred to as Number 16. The broad description is that these trees grow to 14 metres, but it is thought that they can grow larger. The kernel to nut ratio is about 54% and the variety is known as a heavy cropper. We were told to expect nuts after four years. As a variety, Number 16 does not need to be fertilized with other varieties. We were told that this tree could be more viable than the Franquetta and Fernor varieties from France. We had an avenue of cherry trees which we called the "cherry walk" along a public bridlepath which had been planted in 1952 to commemorate the Coronation of Queen Elizabeth II. Over the years the trees had matured, become derelict and were dying. The cherry tree does not have an infinite life and these trees had reached their termination. I decided to cut them down and replace them with a walnut avenue. I was careful to plant them well away from the cherry roots for fear of the walnut being attacked by honey fungus. The soil in this area is sandy to about 45 centimetres, but below that it becomes hard compact sandy clay. I wondered if it was suitable soil for the trees, and thought that, if it was good enough for the cherry, it should be good enough for the walnut.

The first and most noticeable thing about this variety is that it grows much more slowly than its French relation. Walter Fox Allen describes the peculiarities of growth of English walnuts in America. He says that the small trees grow 15 centimetres the first year and the tap root the same. The second year the growth is 30 centimetres and the tap root is the same. For the first three years the root gets established and after that, the tree grows. My experience has been that the American variety produced a small number of leaves off the stem. The second year the growth off the stem was about seven centimetres. Generally, it was noticeable that the growth was slow. That could be a genetic difference, but I suspect that one of the reasons for the lack of growth was to do with the different soils and perhaps roots not finding water in the area I planted. We can have a wet early summer or even a wet July, but as the late summer arrives, the clay just dries out from lack of rain.

One of the ways to maintain moisture is to mulch the trees and we were successful in doing this. We had a manageable number of trees and I obtained a load of horse manure from a kind neighbour. We put a thin layer around the base of the tree, so that it did not touch the base in a circle about half a metre out so that the goodness from the manure would soak down onto the roots. The effect of fertilization is that it improves photosynthesis. It is also thought that chlorophyll levels are improved. Chlorophyll is a green pigment in plants which is essential in photosynthesis. There is no way of measuring the effect of putting manure around the base of the tree, but I believe it helped. The manure rots quite quickly and is gone by the late summer.

Well rotted leaf mould improves the soil by putting humus into the soil and also acts as a mulch. This works well in a garden as the fallen leaves collected from other trees in the autumn can be stored for 18 months and then spread around the foot of the walnut. Because of the quantity of trees we had, the leaf mould option was impossible. While the leaf mould doesn't have much nutritional value, it does add humus to the soil.

A young walnut tree with a mulch of manure at the base.

Growing walnuts from seed

Growing anything from seed is the obvious way to propagate new trees. It is after all, the natural way for regeneration of any tree or plant and the propagation of seeds has been going on since the world began to sustain life. It is easier to transport a seed than to transport a growing plant. It is less vulnerable in its seed state than as a tender young plant. The walnut grows easily from seed and I have done it successfully. My method is to plant the seed in a pot. I used a fairly large pot filled with compost and left it outside to take its chance with the weather. I found that the seed propagated the following spring and there was the young tree. Experts at the University of Minnesota in the United States say that planting in a food can with its top removed and the bottom broken open in the form of a cross is a good way to plant. The can should be burnt first so that it will rust and deteriorate quickly, then filled with about two to four centimetres of earth for the seed to be planted in. The can is placed in the ground covered by about 2.5 centimetres of soil. Why would this method of planting help? Does it make it more difficult for the squirrel or rat indeed to get at the seed?

This seed was found in our kitchen garden in the spring of 2014. To the left is the root which burrows down about five centimetres. Notice that the hard case of the nut is still present. To the right of the hard nut is the beginning of the stem. At this point it is crooked, but it then straightens upwards to go through soil to the surface.

Thomas Pakenham in his book *The Company of Trees* relates that he bought some walnuts from Tesco. He planted some of the nuts in compost and they germinated into saplings. He says that the oil in the nut keeps the nut fertile. He called the area where he planted the trees at Tullinally, "Tesco Corner".

The young sapling is small and weak and needs watering during the summer particularly if the summer is dry or there is a prolonged drought. Probably tap water will do, but I have always preferred to use water from a water butt. The rainwater is pure with none of the chemicals which are used to make water drinkable by humans.

The tree is not big enough to plant out until it is about three to four years old. Experts say that the risk is that when the tree is transplanted, the tap root can be damaged. I use this theory as my argument for the reason why I plant the seed in a large pot. I do not want to replant into a larger pot during the early years of its life. When I plant the sapling out, probably after three years, I do not want to disturb the roots.

It is important to make sure that the new ground is clean and that there is no competition from weeds and that rabbits or hares cannot chew the bark of the delicate tree. The tree needs this important protection and it can be achieved by wire staked around it. Making sure the tree is given the best chance of life lies in the preparation. Experts at the University of Minnesota, Melvin Baughman and Carl Vogt, say that if a commercial plantation is to be planted, weed control should be a priority for three years. In heavy ground, they suggest a chisel plough should be used to give deep cultivation as this will help to break down any pans or underground crust which will inhibit growth. They say that breaking up the soil is important and suggest the use of a rotavator.

The Royal Horticultural Society recommends that 100g per square metre of a fertilizer like Growmore be used in February before the sap rises.

Nature then takes its course and there is a long wait of five to ten years or more for the tree to grow up to maturity and produce a nut harvest. One of the risks of growing walnuts from seed is that the potential nut harvest can be reduced. The same applies to other trees such as apples where growers rely on grafted stock. The reason is said to be that the seed-grown tree is not true to type. Philip Waites, head gardener at Wimpole Hall, says the exception is the Macrocarpa walnut.

When growing from seed, John Evelyn, in his writings, suggested putting a shard of slate under the seed nut, whilst it is growing. This apparently forces the root to spread. That is an interesting tip, possibly one which has been forgotten by many modern gardening experts. Evelyn wrote in the 18th century and this idea is probably not found in forestry textbooks today.

There is, of course, an element of luck in growing trees from seed. Tim Cornish a local historian from Mayfield, East Sussex, researched the story of Anna Bell Irving who lived at a house known as Stone House, which still survives in the middle of the village. Anna wrote a booklet called "Recollection". In it she relates:

> When the palace (previously the Archbishop of Canterbury's summer palace before the Reformation and now Mayfield School) was dismantled, the guest house opposite the middle house (a pub or inn converted from a house, previously owned by Thomas Gresham) was built from the materials in 1730 and there my great-great grandfather Richard Owen Stone took his wife, a Miss Verral of Lewes, East Sussex, to live at the house in 1789. On the first evening when they were walking around the garden, she found a walnut in her pocket which she planted in the ground. That is the large walnut tree hanging over the wall next to the presbytery.

Sadly the tree is not there today, but it did grow and lived 200 years, all from one small seed.

Grafting walnuts

The alternative to growing any tree from seed is to graft species onto rootstock. It is not easy to graft twigs onto a different rootstock. The failure rate may be high, except by the real expert. The principle is that twigs, known as scions are cut from the parent tree. These need to be carefully cut so that they are not damaged. The scions are about the width of a pencil and about 30 centimetres long. The ideal scion has as many buds as possible.

There are several methods. There is the whip graft. The rootstock is cut four centimetres down in order to take the scion which has been shaped to fit the cut in the root stock. The cleft graft is where a cleft is made across the middle of the rootstock using a clefting tool. The rims of either end of the cleft are shaped with a grafting knife. Two scions are fitted to either end of the cleft. The bark graft is performed by cutting a vertical slit in the bark of the rootstock. The scion end is shaped into a five-centimetre wedge. The bark is lifted so that the wedge-shaped scion end can be fitted into the bark. The whole wedge must be embedded in the bark of the root.

In America, there is a practice of grafting English walnut on to black walnut root-stock or onto *Juglans hindsii* which is a native of California. Grafting allows growers to use the indigenous black walnut root stock and quality English walnut scions to give production. That way they get the best quality and quantity of nuts and a valuable timber quality in the burr.

Cuttings

It is quite common to hear of people "taking cuttings" to propagate plants. The idea is that a stem or twig is put in rooting powder in the hope that a root system is formed and growth can begin. I have never tried this with the walnut, but according to the Royal Horticultural Society it does not work.

Stumping

Stumping is a process whereby the stem of the walnut is cut to between five and ten centimetres. It can be done at the time of planting or at a later date. It is in a sense a form of coppicing. Certain trees will thrive on coppicing and become a valuable crop. Chestnut stumps can be hundreds of years old, but the multi stems can be harvested in a 20 to 25 year cycle for stakes or poles. When a stumping is done, regrowth occurs and the aim is to find a single leader which will recreate a new tree. In effect, this is how man can re-start the growing process again.

So why is it necessary to do stumping?

A young walnut, perhaps three to five years old, grown from seed, could suffer from die back as a result of frost. Such damage could render the tree misshapen, in other words not the tulip shape required. There has been research done on stumping at Lount near Ashby-de-la-Zouch by Dr Jo Crook. She says, "In walnut silviculture the most important benefit of stumping is in its promotion of rapid height increment through the early frost sensitive phase of growth." The stumping can be done in the summer or the winter. Once growth has started the leader will become apparent and at that time it is important that other competitive shoots are removed.

2  The walnut grove

I have always been impressed by the way in which fruit orchards and vineyards have, in many well-run properties, a tidiness, elegance and feel of prosperity about them. The work that goes into this vision should not be underestimated. It is probably the case that a vineyard needs the greatest amount of attention. The vines are weak and need much more assistance to protect them against disease and weed competition.

The crop of grapes that the vine produces is far more valuable than the walnut. People have wine on their shopping list regularly in the United Kingdom, but a bag of walnuts on infrequent occasions. In France, and in particular in the Dordogne, the purchase of walnuts would commonly feature on the weekly shopping list. In that region of France the walnut is part of the local diet. In China, where the largest

A well-kept French plantation with short grass and a sprayed strip to keep competition weeds and grasses under control. Not a weed in sight. Walnuts will create a tight canopy which will suppress undergrowth.

quantity of walnuts in the world is grown, local rural populations consume these readily available and nutritious nuts as part of their diet. This is important because it shows that people in rural communities are able to feed themselves on local produce. They can eat and sell the surplus.

One of the questions that occurred to me was whether a nurse crop for the walnut would help to push the growth of the walnut grove? Nurse crops have historically been used in forests. Trees such as oak and beech benefit from a nurse crop. Hazel as a nurse crop within the oak planting is known to draw up the oak. Capability Brown planted larch amongst his clumps of oak, elm and beech. The shade from the nurse crop has the benefit of preventing side growth of branches. The side growth produces knots in the timber when felled. The scientists at Paradise wood in Oxfordshire have looked at this idea of nurse crops within a walnut grove for timber production. The aim with growing walnut trees is to create a wine glass or tulip shape and therefore a number of low branches are desirable. The walnut tree grows slowly and hates shade. This evidence is another good reason to have no other trees, nurse crop or vegetation growing near them.

Any work with trees means management and care to ensure the best growing environment. It is said that some French farmers, particularly in the Grenoble area, turned to walnut groves as they needed less work than a vineyard. In other commercial tree operations such as apple orchards in United Kingdom's counties like Kent, the husbandry technique is to kill the grass under the tree in order to keep the base of the tree clear of vegetation. This vegetation will take up water before it gets to the

Walnuts are seen here growing on the edge of fields of maize.

root of the tree. The same applies to the walnut, but when harvesting nuts, the clear ground makes it easier to collect those which fall to the ground. Killing the vegetation under the tree, so that the surface is almost brown earth is one way. Perhaps a more environmentally friendly way is to cut the grass as one would a lawn. A severe cut makes the nuts easy to see as they fall and if a machine is being used to gather them there is less material for it to pick up or cause a blockage.

Cutting the rest of the walnut grove on a regular basis encourages the growth of grass, enables good access and discourages the growth of weeds which are unattractive and take up more water than grass. On our farm my targets are dock, nettle, thistle and ragwort (that yellow daisy-like plant seen along the roadsides). I found that by cutting every two to three weeks soon after planting and through subsequent seasons, these weeds are weakened and eventually die off. I have also found that cutting early in the year, as the spring grass comes, is good because it makes cutting during the rest of the year easier. During the spring, the grass grows the fastest and therefore needs more cutting. Gardeners know all about this as they work their lawn mowers.

In France the farmers are often more conscious of the space they have available than we are in the United Kingdom. The Napoleon Code says that land is distributed equally to all children following a death. Sometimes a patch of trees in a field might belong to one person and the surrounding crop belongs to a brother or sister. In the Dordogne, white crops like wheat, maize and barley grow very close to and amongst young trees. In other very rural parts of this agricultural area, vegetables such as

These vines are growing under a smattering of walnuts in the Dordogne.

broad beans and lettuces are found growing. I have even seen strips of asparagus between rows of trees.

The University of Missouri Agroforestry department recommends that for the best cropping of black walnuts, there is 12 to 15-metre spacing. This will allow inter-cropping and agronomic crops.

3

Farm animals to benefit the trees

I did consider other ways to keep the walnut grove clean and tidy, as well as providing a bit of income. Nature does not do monoculture, as I have heard said. Between the alleys of trees growth of other plants will appear. Grass, weeds or, with real neglect, brambles and saplings. In France, the farmers may grow other crops or simply keep the alleys cut. The obvious alternative to monoculture in the UK is to combine trees with livestock or crops to create Agroforestry. It could be cheaper to graze than it is to cut grass. Time and fuel costs being the main contributors to cost. The alternative is spraying, but that too is expensive and in time could upset the soil structure.

Grazing sheep is a possibility. They would certainly keep the ground clear and the grass short. The trees give shelter to the livestock. Lambing is a critical time for the sheep farmer and shelter plays its part in the husbandry and management of this important event. After that, the ewes would benefit in the summer from the shade provided by the mature trees. There would also be that satisfying feeling that the land was being used to its maximum capacity. The other thought I had was the provision of nitrogen from the ewe's manure in the soil. The possible benefit here is that the manure will not only improve the tree growth and nut production, but provide fertile ground for the ewes in the future.

Nitrogen could come from the planting of clover under the trees and in the grass sward surrounding them. There are two nitrogen-fixing clovers: red and white. Red clover, not widely grown these days, is an efficient crop for providing nitrogen to the soil. With the trees widely spaced, the red clover would provide grazing. The disadvantage with the red clover is that it lasts only two years before it needs to be replaced. When the trees grow the ground vegetation will disappear and the clover with it. The white clover is short, will not grow too tall and provides a little nitrogen to the trees but will be mixed with other grasses.

From a management point of view, I think it best practice to graze intensively for a period and then turn out the sheep. In order to get this right, the grazing must not interfere with the walnut cropping. March to May grazing would be one period to consider even though lambing could be in progress. Another time would be in late October when the nut harvest is over.

Our farm is not big enough to have an economic number of ewes and the grazing would have to be done with the co-operation of a local farmer. That is not a problem in itself, but I would not have control of moving the sheep on or off the land as I might wish. Grazing is normally agreed for a season and the tenant has the right to graze as he sees fit. In the summer before shearing the wool off their backs, some sheep will rub off their fleece. In order to do this, there is nothing nicer than a tree. The ewe needs a tree which is solid and preferably with a rough or uneven bark. The effect is that this rubbing damages the bark. That means individual protection for the trees as well as fencing the outside of the orchard to keep the sheep in. The young tree is also at risk as sheep can develop a taste for the bark. They can be grazing peacefully one day and the next start to strip the bark. The prospect of these problems and the work to go with it, seemed to me to be a task too far. It was not only the disadvantage from a labour point of view, but also from the cost of doing it.

Grazing cattle also has the benefit that it keeps the grass down. Cattle graze in a different way to sheep, eating almost anything in the field. Sheep will nibble and be selective in their diet. Because of the cattle's grazing habits, it is fair to expect them not only to graze around the tree but also the leaves on the tree itself. This is particularly the case if the tree is fairly young or has low branches. They will also rub up against the tree and the larger cows will brush under the tree breaking branches. There is a high chance that cattle will damage the tree, particularly the outer branches so that they bleed sap. If you put individual guards around the tree, cattle will rub up against them and break them. My advice is, that until the plantation is well established, do not allow any cattle in to keep the grass short.

A less effective, but slightly more practical, idea is to have free-range turkeys to fatten up, probably for Christmas. The damage they could do to the trees would be minimal. The life cycle of the turkey from birth to slaughter is 22 weeks and to really finish them it could be 26 weeks. The problem with this half-year cycle is that the birds may be on the ground during the walnut harvest.

Chickens on a free-range basis could be an option. The difficulty is to manage the project in a cost-effective way. Small numbers do not make economic sense. On the other hand, with just a few hens, selling eggs to local shops or consuming them in the family and friends circle may be satisfying.

What about geese? They are tougher than the turkey and are fierce. They can beat off a fox and may well survive harsher conditions. The geese will fertilize the ground, but could also sour it. There is the problem of harvesting with birds on the ground. On a positive note, for the buyer, a goose at Christmas is appealing.

The possibility of poultry came from my theory that the orchard had to be secured against deer with the use of electric fences. It would not be too difficult to make the compound fox proof and prevent the turkeys, chickens or geese escaping. I saw it as something they would do in France or Italy, albeit in a less intensive free-range way than I was intending. From a husbandry point of view, I took the view that the trees would provide some shelter particularly for the hens, but temporary additional housing shelters would be required at the onset of winter. Poultry in woodland is not a common sight in the UK, but there are people who do it in quite wild places, like Northumberland, and use the woodland environment as a selling point. Chicken manure has ammonia in it. The trees will take up the ammonia and therefore there is an environmental benefit.

So why did I not jump at this idea? Whilst I have had a little experience of poultry, I would not pretend to be an expert. Turkeys are quite delicate, they can easily succumb to disease and the risk of financial loss to the unwary is quite likely. My knowledge of ovines (sheep) and bovines (cattle) is much greater. There was also the question of labour. It would be quite intensive for this kind of project and I would have to employ labour with adequate knowledge to do it. The labour would be on a part time basis and would need to be found year-by-year in all probability.

The idea of pigs crossed my mind. The problem with pigs is that they forage and ruck up the land and I suspect would cause damage to the trees. Again, there is also the question of labour.

4 The enemies of the walnut tree

Deer

The damage deer do stems from a natural instinct for a young stag to want to be rid of its velvet. In the deer's annual life cycle, the stags shed their fine antlers each year and renew them. The new antler has a covering of fur called a velvet. In some countries the velvet is much prized for medicinal reasons, but not, to my knowledge, in the UK. In time the velvet begins to die or rot which seems to cause irritation and the beast needs to shed it by scraping or rubbing it off. Putting myself in the mind of this beast, I could think of nothing better than a two-metre whippy walnut sapling to scrape off an irritation which in human terms must be like a very bad itch. This activity is called "fraying", according to Richard Prior writing in the *Shooting Times*, and evidence of it is seen as vertical or diagonal scarring of the bark. I have noticed this fraying activity in January or February by roe deer and in August or September by fallow deer. Fortunately we do not have muntjak, an animal first introduced to the United Kingdom in 1925, to Woburn Abbey from South Asia. They are indigenous to India, Sri Lanka, Myanmar (Burma), the Indonesian Islands, Taiwan and Southern China. The translation of the name is the barking deer and with that name they can cause considerable damage. The ten minutes or longer it takes to complete this fraying task, will kill the tree because it will wear off the bark. The bark is vital for the tree's existence and it will die without it, since it protects the tree and assists in water rising to the leaves to photosynthesise.

In order to minimise damage on my farm, something had to be done. I built a stock fence around the grove and then added two strands of electric-fence wire, and used a battery-fencing unit which ran off a car battery. It certainly worked, but not as well as it would have done if I had used a unit fed off the electricity mains. I will admit, I have a fear of fire and didn't like the idea of the mains attachment, particularly during periods when I was away. In order to connect a mains fence safely, proper wiring and connections have to be installed. I was told firmly that the use of extension leads with ordinary plugs was not safe. My rejection of the mains option may have been an error, but that was how I preferred to run the deer fence. There was another problem with electric fencing and that was preventing it from shorting when vegetation grows over it. I found it extraordinary how often spraying or cutting was required over the course of a summer. One of the ways round the problem of vegetation growth is to build a more solid fence with sheep wire at the bottom and two electric wires above it.

The deer problem was one I overcame in another area by manufacturing my own tree guards. I was lucky that I had a supply of chestnut from my own woods which I was able to cut and use as the main posts around the trees. I knocked four posts into the ground so that the tops were two metres above ground level. We used a two-man post driver to complete this task and firmed up the structure by nailing seven by five centimetre tanalized timber at the top. It was then just a matter of putting some sheep wire, a metre and a half high, around the posts. It was necessary to spray weed killer around the base of the tree inside the guard once or twice a year, but at least there was not an electric fence to maintain. The disadvantage with this method was that cattle could reach the lower branches and graze them if short of grass.

The rabbit and hare

Another pest is the rabbit. Rabbits will take delight in biting the bark off the bottom of the tree. The bark is part of their diet and they certainly have teeth that are capable of stripping the tree. Again this can be fatal for the tree. Plastic spiral rabbit guards that expand as the tree grows are the simplest and most effective solution. The hare is also guilty of this crime. But for the financial damage this romantic and mysterious animal can cause, I have a great affection for it and wish it no harm.

The woodpecker

Another threat we were warned about (through Manse Organics) was that of wood-peckers causing damage to twigs. There are three types of woodpecker in our part of Sussex, particularly the colourful green woodpecker. This bird is a welcome sight, when glimpsed in flight. You spot this flash of green and red as it darts through the air. We have not seen any sign of such damage during the years we have had the walnuts.

The squirrel

The squirrel presents a different problem and one which needs careful management as there are animal welfare issues involved. The squirrel to which I refer is the grey squirrel which was imported from America. The grey squirrel arrived in Denbigh-shire in 1828 where it was kept in captivity. The population was released in 1876 to spread and expand in prolific numbers over much of the country. The much-loved red squirrel is half the size of the grey and has caught the imagination of the public. Its popularity has led to schemes encouraging its reintroduction. Experts say that the

Abraham Mignon, 1640–79. Detail of *Fruit Still Life with Squirrel and Goldfinch*, c.1668.

grey is responsible for the decline in the red squirrel population as it carries a virus lethal to the red. The grey is responsible for damage to many species of trees that can be extensive. I grew a stand of sweet chestnut after the Great Storm in 1987. After ten years I had to fell most of them as the squirrels had stripped the bark out of the tops of most of the trees. The damage had the effect of halting growth. Fortunately, as the chestnut is one that can be "coppiced" (cut down to the lowest part of the trunk) there was regrowth. If this happened to a walnut, it would not regrow into an acceptable shaped tree and to take the tree out would be best.

It is not easy to balance public opinion on the charm of the grey with the pest it has become. The size of the grey squirrel, that bushy tail, its acrobatic climbing ability, its cautious hopping on the ground, always looking for danger, and that soft-looking fur, gives the image of a lovable animal that can do no harm. This is particularly visible in St James's Park, London where the squirrels are fed from the hand. Sadly, that is not the case and this little animal can inflict heavy economic loss together with environmental damage. It is not surprising that gardeners, farmers and foresters would like to see their numbers kept down. The other disadvantage with the squirrel is that it breeds twice a year. That can mean a very large population develops in a relatively small area in a short time. Yet, when there is a pressing need to reduce animal species for health or economic reasons, the authorities will act. They did so when the badger was accused of spreading tuberculosis to cattle. On the coast and in estuaries, the seal has often been threatened with a cull for its voracious fish appetite and the effect it has on salmon stocks. The squirrel does not have a known health risk to humans, does not appear to create a meaningful economic disadvantage and is unlikely to come to the farming and environment authorities' attention for eradication or reduction.

Laudably, certain farmers and estate owners have resolved to reintroduce the lovable red squirrel. There is a project in Cornwall and a successful one in Anglesey. They also thrive in Jersey.

I hear from people who have walnut trees in their gardens that, just before harvest, the crop they would have had has been entirely removed by this predator. The squirrel is clever and can detect when the nuts are ripe. The days get shorter in the autumn and perhaps as they feel the onset of winter they need to hoard food. They usually take the nuts and hide them in shallow holes some distance from the tree. It is some feat to do this, if one considers the size of the nut and the ability of the squirrel to carry an object which is big in proportion to the size of its mouth. It sounds an odd thing to say, but it seems a psychological certainty that this little intelligent animal will steal a smaller nut which is easier to carry if given the chance. Chestnuts and hazelnuts would be an obvious smaller alternative if they are in the vicinity, but it is not likely that a chestnut or nut-producing hazel crop will be close at hand to most gardens. Therefore the walnut becomes the thief's choice.

Squirrels will take a green nut and strip off the outer rind. We find piles of them under the trees alongside nutshells where they have broken in to find the sweet nut. It is distressing, but I feel satisfied that when we pick ripe nuts at least they will not be taken by the squirrel.

I had a live trap, which was a cage with a trap door and baited with peanuts, on the edge of my grove, which I kept a daily eye on. I began to notice rind under the trap. It became clear, I had a tree rat squirrel with a sense of humour at my expense. He even had the cheek to spring the trap without being in it.

On a *Gardeners Question Time* programme (August 23rd, 2013), one of the panel suggested that a solution would be to put a layer of mulch, which could be leaf mould, under the walnut tree so that the squirrel has a ready-made place to bury the nut for future consumption. When the tree has shed most or all of its nuts, just rake up the mulch and retrieve the nuts. A simple idea which is practical, but I can see a disadvantage with it. The mulch would dampen the nuts and they would need quite a lot of drying. A laborious way to get around this problem would be to search the mulch each day.

According to a letter to *The Field* (November 2014 from James Edgar), the squirrel can be a tasty meal for a fox if it is stalked on the ground following the theft of nuts from a walnut tree. It was also mentioned in the letter that there is a possibility that the same squirrel had been badly ragged by a pair of magpies who did not like the raid on the tree. Their disapproval was overcome by the probable stubbornness and determination of the squirrel.

There is a natural predator which is a killing machine and is keen to dine on a squirrel, if it can catch one. That animal is the pine marten, a shy mammal which is found in Europe and in the United Kingdom mostly in Scotland, but also parts of England. It is a mammal that is growing in numbers. The pine marten hides during

A squirrel making
away with its prize.

the day and hunts at night on the ground and, most importantly it can climb trees. It is quick enough to catch a squirrel. This animal is not found in the south yet where most of the walnuts are. It does not pose so much of a threat to the red squirrel which is lighter than the grey and can escape the pine marten over the thinner branches. There are many people who believe that the pine marten should not be introduced to prospective red squirrel areas because of the risk of other animals domestic or otherwise being killed.

There are some types of trap which are described as "humane killers", but others that are designed just to trap and not kill. From my own experience, it is not easy to catch squirrels alive. It helps to watch the routes the squirrel takes. They seem to have daily habits and to visit the same places and in particular the same trees. They know their territory. Without doubt, the squirrel has extraordinary curiosity and is intrigued by anything new. I have found that, if traps are put in different places and are not set for long periods, the catch rate is better. The squirrel is a very capable jumper. I watched and learnt not to underestimate their ability to make a vertical leap of a metre and a bit onto a low hanging branch. I have seen a squirrel leaping on to a low branch rather than climbing the trunk. The low branch needs to be cut to at least a metre above the ground in order to encourage the climbing squirrel up the trunk where a trap should be placed. Equally I have seen squirrels leap from one tree to another over quite a distance, particularly if the jump is from a higher branch to a lower branch. Experience has shown that this is a good reason for setting the planting distance at five to seven metres in a garden or a walnut grove.

The dispatch, following the entrapment of the squirrel in a live trap, is an important animal welfare consideration. It is against the law to release vermin after entrapment. A quick dispatch without time for the animal to suffer is the ideal.

Shooting, in the field without trapping is a method which is an obvious solution. The act of shooting even if the squirrel is wounded is legal, but it is not easy to do this. The best weapon is an air rifle, but a modern high-powered version. In order to fire a shot in the right place to kill the squirrel instantly, a well-zeroed telescopic sight is required. You have to be able to stalk close to the target and this needs patience and a good idea of where the squirrel is likely to be found. I was given some advice which suggested the best way to stalk the squirrel is to lay a bait area or feeding station so that the squirrel comes to feed at the same spot regularly.

Squirrels live in dreys which are a kind of nest built high in a suitable tree. An experienced person will have no problem in spotting them. The first of two mating seasons is in the early spring and the young are born in the drey. This is the time when there is some certainty where the female will be when she is caring for her young before they set out on their independent lives. It is also when they can be controlled by shooting.

There are ways that the tree can be protected if the squirrel cannot be controlled. The trunk of the tree is an obvious point of entry for the acrobatic squirrel. The squirrel has a remarkable anatomic characteristic. It can twist its ankles into positions so that it cannot only climb a tree, but climb down a tree. In effect it can turn its ankle 180 degrees so that it can grip the tree as it makes a descent, often eating upside down. There is no other rodent like it and this single ability makes this tree rat such an effective thief.

A tin casing around the bark of a metre high may prevent the squirrel from climbing a tree and using its anatomical climbing ankle, according to Clive Simms in his *Nutshell Guide to Growing Walnuts*.

5  Walnut varieties

The genus of the walnut is *Juglans*. The family is Juglandaceae. The Hillier *Manual of Trees and Shrubs* lists twenty varieties, including the common walnut, *Juglans regia*, and the black walnut, *Juglans nigra*. There are two other related trees: *Pterocarya*, the wingnut and *Carya*, the hickory

These genera of tree are found in most temperate areas of the world, including Europe, The United States of America, Asia and New Zealand.

The wingnut

The leaves are typically pinnate, 30 to 60 centimetres long and toothed. The tree is wide in its growth.

A wingnut at Albury in Surrey. This example is *Pterocarya fraxinifolia* and it comes from the Caucasus.

The butternut

This tree is a native of eastern North America and was brought to Great Britain in the 17th century. It is rare to find them in this country. There are 7 to 17 leaflets. The fruit hangs down in clusters of 5 and a mature tree could be 15 to 30 metres high.

The trunk and the leaf of the Butternut or *Juglans cinerea* at Thorp Perrow.

The common walnut *Juglans regia*

The migration

The common walnut with a regal name, *Juglans regia*, has many other names including the King's fruit, old world walnut, Jupiter's acorn and the soft-shelled walnut. It is said to have come from Persia, which is the origin of its other name, the Persian walnut. The migration of the walnut to the west is an extraordinary story. As evidence of this migration, there are walnuts in the Lebanon which were described by Sir Joseph Hooker, a noted 19th-century explorer and director of Kew Gardens. The tree would have been planted from seed transported overland and by sea at that time.

Archaeologists found a 5th-century chestnut and oak boat in the port of Yencapi near the modern day Istanbul, in Turkey. Inside the boat, believed to have been wrecked in a storm, they found walnut seeds. The port was important as it was located where the Mediterranean and Adriatic meet the Black Sea and the mouth of the Danube. Nuts were transported to Greece in about 100 BC. The Greeks selected nuts from the better trees to improve their groves and it is known that they were making walnut oil in the 4th century. According to Greek mythology, it is said that Dionysus changed Karya, the king of Laconia's daughter, into a walnut tree. He loved her so much that he felt he needed to transform her into a beautiful tree. It is notable that the Greek word for walnut is *Karya*, named after this important lady.

The *Juglans regia* is found in Romania and to the north in the Carpathian Mountains up to Poland. On its migration west, it was taken to Italy where the *Juglans* name came from the Latin. It is possible the nuts first made landfall in Italy in the south near Sorrento and to the volcanic area of Vesuvius. The next migration was west again into France, Spain and Portugal.

The mosaic shown below was taken from the dining room of the Emperor Hadrian's villa on the Aventine hill in Rome. It dates from the 2nd century BC. It was created by the artist Heraclitus and depicts the aftermath of a banquet. Items in the collection of debris include chicken legs and lobster claws. There is also humour in this mosaic with the mouse stealing the walnut. The mosaic has thankfully been taken to the Vatican Museum in Rome where it can be seen by everybody.

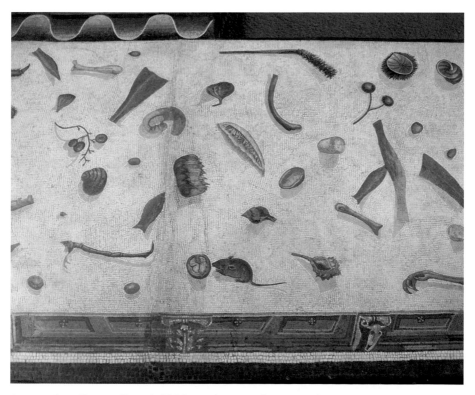

Amongst this collection of household debris is the image of a mouse and an open walnut.

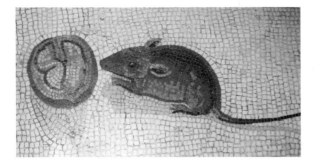

A close-up of the mouse and the open walnut in the Heraclitus mosaic at the Vatican Museum. The date of this mosaic indicates when the walnut had reached Italy. It must be that during the reign of Hadrian, the nut was common in Italy.

The nut kept moving further into Europe with the Roman army and was introduced to Britain during Roman times. In the 16th century seeds and grafts were brought in from Europe to plant new trees. The landed classes of England were fascinated by the growing of fruit and nuts. Maggie Campbell-Culver in her book *A Passion for Trees: A Legacy of John Evelyn* says that Evelyn reported a number of walnuts in the Surrey area near Leatherhead and Carshalton. She mentions that a Peter Mason of Isleworth had no less than 2,000 walnuts in his collection in 1730.

Another route into Britain came in the 19th century. It is thought some seed was brought to East Anglia when troops returned from the Peninsular war. These men, from farming stock, planted them in gardens and field edges.

Juglans regia are found to the east of Persia and as far as India and Myanmar. Kyrgyzstan to the east of the Caspian Sea and on the Silk Route has 230,000 acres of walnut trees. The type of nuts vary and, as the tree will grow at altitudes over 1,000 metres, Gabriel Hemery and Sarah Simblet say in the *New Sylva* that the mountain ranges of the Himalayas, Hindu Kush, Pamis and Tian Shuan have preserved a uniqueness in the tree.

The tree

The common walnut is described as slow growing up to 30 metres, with a rounded head of spreading branches. It is valuable for its nuts, timber and presence. It will survive at temperatures as low as minus 15 degrees, but a cold snap in the early spring can have a devastating effect on the crop for the forthcoming summer and the tree's growth can be severely impeded. The bark is grey and smooth, but will become fissured later in the tree's life. The leaves give off an acrid but not unpleasant scent. They are leathery to the touch and are bronze when they break in the spring. The leaf then turns a dark green in the later summer.

It is reproduction that creates the nut. Many fruit trees rely upon bees or other insects to transport the pollen to the bud. This is not the case with this tree. The walnut is monoecious, like many other trees, which means that the tree produces both a male catkin and a female flower. In some varieties the catkin will appear

Juglans regia

Joseph Hooker, 1814.
This illustration
shows elements of the
walnut – the catkin,
the leaf, the flower,
the new growth and
the nut in half.

before the leaf. The catkins are found slightly down the twig whose growth was created in the previous year. The all-important bud at the top end of the twig is the female flower. It is essential that there is a slight wind at the right time so that pollination can take place. I use the word slight because any really high wind will spread the pollen but, wider and further than ideal because the pollen has to be caught by the bud. The ideal spring for the walnut is one where there is an abundance of catkins and the air is filled with pollen on the right days. Unfortunately, the pollen migration at the right time does not always happen in our climate. The British spring seems to vary from one year to another. The length of winter varies from year to year and this affects the growth of the bud and the catkins. It is important that there are different varieties in the plantation as this improves pollination.

The catkin has a calyx of five or six scales surrounding thirty-six stamens. The catkin is yellow and green in colour and is very distinctive. They can be 5 to 15 centimetres long depending on the variety. The catkins appear from the middle of April to the middle of May. The pollen is fertile for a period of five to six days and it is crucial that the bud is fertile at the same time.

The flower grows at the end of the branch or at the terminal bud in the old varieties. The walnut specialists in the Dordogne are trying to breed trees which produce flower buds further up the branch (that is to say towards the trunk) in order to increase the crop. The flower is not as obvious as the catkin. Many people see the catkin and think that this is the flower. The flower is green and may have a cone type shape. It appears between the middle of April until the first week of June. The flower can be fertilized for a period of eight to fourteen days. That is why the pollen, which is only fertile for five to eight days, must coincide with the flower's fertility. Some trees are heterodichogamous, that is to say that the male pollen isn't fertile at the same time as the flower or vice versa. There is no scent from the flower. The pollen is blown onto the flower. It can fertilise its own tree or a tree up to 200 metres away. My walnut grove faces south and the prevailing wind is from the south-west. We notice that the trees that are further downwind are often more prolific in nuts than those that are near the southern edge. We also notice that the trees which are near a copse at the bottom of the hill and are therefore sheltered from the prevailing wind do not produce so many nuts.

When the trees are young, it is obvious that there are fewer twigs and therefore fewer catkins. The fine dust, unseen by the naked eye, is vital to the success of the crop. The quantity of pollen makes all the difference. In theory, and over time, as the tree grows so does the volume of the crop. In practice I think it is a combination of this growth and the weather that determines how many nuts will appear on the tree.

The flower is small, very delicate and is easily destroyed by late frosts. Such an event could destroy the crop for the year. In 1618, Richard Stapeley records frost and snow in May, this record was backed up by the diarist, John Evelyn. Few nuts would have been produced with weather like this. Disaster struck in France, when

Catkins ripening for pollination in the spring on a *Juglans regia*. The black walnut's catkin is different in that it hangs down.

the temperature in the early part of April 1991 fell below freezing in the Dordogne. It was calculated that 80% of the nut crop was lost. It is obvious what has happened as the young leaves are blackened and die off. The tree itself does not die because new leaves form a few weeks later. In the UK there is a variety called Broadview which has some resistance to frost damage.

I have noticed that when a tree is young and there are no catkins or a chance of nuts, the tree comes into leaf earlier than one which is trying to produce nuts. It seems there are a set of priorities. The young tree wants to grow and get the sunlight. The older tree is looking for reproduction and wants to create its offspring walnuts. After that, the priority becomes growth into a fine tree.

The catkin's use comes to an end when the pollen has flown. It blackens and eventually falls to the ground. In fact there are so many of them in a good year that there is almost a carpet of them beneath the tree. It takes a short time for them to disappear altogether. As the spring progresses, the light is better, the days are longer and the air warmer, the leaves break out on the twigs. They are a golden bronze colour when they first appear. They look striking against other trees which green up quickly, but in a short time the leaf turns green as it matures and blends in with its surroundings. In the late summer, a bronze-coloured leaf appears, a normal development as the tree is still growing. As these shoots appear, so do new leaves. The shoots themselves are green when they first emerge, but later become brown, as one would expect.

Once the bud has been fertilised, the flower begins to form and in time it will turn into the nut.

The nuts become noticeable when they are about the size of a pea with a small leafy tassel. The tassel disappears later on and should be gone before the nut is considered ready for pickling. This is the first harvesting time and it is the time when the lower nuts can be taken with ease. At this point, the tree relies upon a normal summer weather pattern with warmth, sunshine and moderate gentle rain. A combination vital to the tree in order to enable it to ripen the nuts and prepare itself for winter.

Very young walnuts are shown here with tassels. Later the tassel will fall off.

The common walnut in Britain and its history

The common walnut is mainly found in France, Spain, England, Australia, Morocco, Circassia, the Balkans, Armenia, Azerbaijan, Turkmenistan, California and Turkey. In Britain the tree has not anything like that popularity for commercial growing, being more for decoration rather than large scale commercial use. There are some good examples of the *regia* in plantations as well as individual examples.

In Maggie Campbell-Culver's book *The Origins of Plants* we are told that the earliest recorded walnut grove in England was planted at Wilstrop in Yorkshire in 1498. Apparently the local population were not in favour of the planting and cut the trees down. The reason for their disquiet is not clear.

There are historic records of walnuts planted at Lyveden. In 1597 Thomas Tresham who owned the manor and designed the garden, was in Ely prison for his Catholic views. He had a passion for fruit and nut trees, and as part of his scheme he ordered warden pears, medlars, black cherries and walnuts from his nurseryman, Andrewes. The question is where did Andrewes get the plants? I suspect that they came from Europe.

In his book *The Stocks Held by Early Nurseries*, John Harvey describes tree stocks held by some famous nurseries. Before 1660 there were no recorded nurseries. Captain Garle, (1621–1685), had a nursery near Spitalfields and became the King's Gardener in 1673. George Reckless started his nursery in Hoxton in 1665, declining in 1691. The Brompton Park Nursery which was on the site of the Brompton Oratory was started by Looker (senior partner) and the Queen's gardeners Moses Cooke, John Field and George London. It was George London who brought in Henry Wise in 1687. It was a 50-acre site and Wise sent trees to estates like Blenheim. William Cox Senior was a nurseryman at Kew and his son took over from him. Peter Mason was another nurseryman and was based in Isleworth. John Berry lived in Tytherington which was 16 kilometres north of Bristol. Berry's story is a different one. It seems that between 1714 to 1721 there was a stock of 5,000 walnut trees.

Brompton Park Nursery had in summary 1,200 elms, 900 hornbeams, 600 horse chestnuts, 450 sycamores, 440 limes and 200 walnuts. These are mostly forest trees and although the walnuts were not the most popular they had a good showing.

Berry sold apples, cherries and pear treess, but he had a stock of 100 walnuts.

Mason had a larger collection of trees. Many of them were grown for forests. His stock included 8,000 English elms, 6,000 Dutch elms, 4,000 horse chestnuts, 4,000 beech, 3,600 limes, 3,250 hollies and 2,000 walnuts.

The locations of these nurseries are interesting. They are all within easy reach of ports at London and Bristol. Most of the plants at that time would have been imported either as seeds or saplings from Europe, in the case of the common walnut, or America, in the case of the black walnut. The ships would not have provided ideal conditions for young plants with a combination of heat, poor light, a lack of fresh water and cramped holds in the ships. There was the risk of storms, wrecking and piracy, but more importantly there was a need for someone to look after the stock. Once the stocks were brought ashore the conditions needed to be changed quickly in order to revive them.

Planting on a grand scale in this country has not been done for a long time and is not common these days. Ashby-de-la-Zouch is in the heart of the national forest. 70,000 trees have been planted of which there are 13,000 walnuts. The total area is 320 square kilometres and the walnut part of the plantation is near Lount. The trees cover an area of 28 hectares. The land is reclaimed from coalmines and mineral workings together with some farmland converted to forest. The walnut growing is an initiative promoted by Jaguar, the car manufacturer, who have used walnut veneer for their dashboards since the cars were first manufactured. The planting was done to create a plantation and apart from the timber there is the aesthetic and conservation value. These walnuts could be in the forest for hundreds of years, but ready for their use as timber in 50 to 75 years. The success of the forest as a walnut production centre depends on how the managers consider the continuity of supply. This is something that has been done successfully in California. Management is not just for the nut production but for the timber as well. Veneer manufacturers look for continuity as well as quality.

Recent plantings of walnut can be found elsewhere. A golfing enthusiast who is a member of Walton Heath golf club, a timber expert and tree enthusiast, has planted 300 walnuts on the edge of the course. They have been planted as a ring around other trees in an environmental project.

The beauty of the walnut can also be demonstrated when it is planted as an individual tree. Hugh Johnson in his book *The International Book of Trees* says, "What makes a man proud and happy but having a walnut on his land. There can scarcely be a tree that puts on less of a show."

The walnut looks better planted with plenty of space around it so that the viewer can see the real shape of these superb trees. The landscape of the garden is in the hands of the designer or the owner. It is probable that the owner may just find a spot in the garden but others, and I would hope the designer, will find a spot to show off this majestic tree. Gardens are a British passion as they are in other parts of the world. There are many reasons to plant a garden, but the main ones may be in order to improve the look of a house, to provide pleasure and pride, to give colour, to entertain, to provide food or medicine and perhaps to provide peace, shade and tranquillity.

Landscaping became popular in the 18th century. Capability Brown created superb parks for some 60 great houses up and down England. He did not plant walnuts very often even though common walnut seed had been brought from France and black walnut seed from the United States. He favoured forest trees. His choice included the indigenous beech, oak and elm, but his aim was to plan for beauty, shade, views and a natural landscape. He was however aware of shape and could see the benefit of clumps of trees in the landscape. At Corsham Court near Chippenham, Brown was commissioned by Paul Methuen, a wealthy clothier, to create a pastoral scene (according to the Corsham Court website). A plan was drawn up in 1761 to achieve the transformation. Apart from the planting of tree belts around the boundaries, an elm avenue was kept to the north and south of the house with supplementary trees planted. There was an avenue to the east which was severely thinned to make way for picturesquely grouped trees comprising mainly of oak, ash and walnut. Another Brown landscape was at Youngsbury near Wadesmill, Hertfordshire. The features said to reflect Brown's plans of 1770 today are mature trees that we are told include oak, walnut, horse chestnut and sweet chestnut. Youngsbury was built by David Poole in 1747. There were 97 acres of land in the estate. David Barclay, a Quaker banker, bought the estate in 1769, and was the instigator of the Brown plan.

Avenues providing walks or rides were part of Brown's planning but I have seen no reference to any walnut planting as an avenue. Fashions for planting change and some of Capability Brown's plans have been changed. The garden at Wimpole Hall, in Cambridgeshire, was planned by Brown and is home of the National Walnut Collection. The remains of a walnut avenue said to have been planted in Victorian times survives but there are now gaps where trees have died or been felled. The gardeners at Wimpole have tried to restore the avenue by filling in the gaps.

Wimpole Hall, near Cambridge. The park was designed and built by Capability Brown, not known for his interest in walnuts. Dispite this history, it is now the home of the National Walnut Collection. This idea of a collection is not something that is found in France, but the British seem to like centres of excellence.

The remains of the walnut avenue at Wimpole Hall.

The walnut in a garden

Were there walnut trees in the Garden of Eden? The Garden is said to have been located somewhere in Syria or Iran. The Old Testament says that all the trees of the earth grew there. That seems improbable as many trees would not have liked the climate in the garden. It is thought that trees such as apples had their origin in the area where the garden was supposed to be. If that is right, then why should not walnuts have grown there?

The word paradise is significant because it is an old English word meaning a place of great beauty. There is also the Latin word *paradisus* from which the English word is derived. The online etymology dictionary describes the word paradise as a Greek word originally used for an orchard or hunting park in Persia. In other words, this is a description of the Garden of Eden. The biblical garden is in the vicinity of where the *Juglans regia* is said to have originated. Not far away in Georgia and eastern Turkey there are forests which is why there could have been walnuts in this famous but extinct garden.

Another word we see from time to time is Arcadia. The word has a Greek origin with references to pastoralism and harmony. Adam Nicolson describes the creation of Arcadia as the ambition for peace and tranquillity in his book *Arcadia*. The book he wrote is about the house and grounds of Wilton House near Salisbury, where he suggests that the Earls of Pembroke constructed the gardens in a manner whereby they, as national leaders, could seek peace away from the affairs of state. The idea of the gentle walk amongst fine trees and shrubs in order to distract a person from their worries, concerns or decisions, was fashionable in the 18th century. Wilton provided a theme for several gardens and landscape designers such as Repton. The other image for a garden is purely for show. The idea of the need to show off wealth, power and taste appealed to many of the rich. Wimpole Hall has this Arcadian image with the stroll amongst its wonderful collection of walnut trees. Lord Petre at Thorndon Hall planted many walnuts and again he may have found the show, beauty and peace in their company appealing.

One such tree is *Juglans Linnaeus*. The upper branches are smooth and leaves are eight to twelve inches long. The leaflets can be seven to eleven in number. Another popular walnut variety is called Broadview. W.J. Bean mentions others in the fourth edition of *Trees and Shrubs Hardy in the British Isles* (1924).

> Bartenani: The nuts are almond shaped.
> Heterophyllus: The leaves are long have irregular lobes.
> Laciniata: It has lacerated leaves.
> Maxima: The nuts are twice the size of normal nuts, but do not store well.
> Monophylum: A tree with one large leaflet and two small ones.
> Pendenta: The branches are pendular.
> Racemosa: It has clusters of 10 to 15 nuts.

Allan Hunt of the Walnut Tree Company picks out Rubra or Red Danube. This tree has red kernels. It is a rare tree and is suitable for gardens. Other varieties include:

> Lara: Good in hedge rows.
> Fernette: Good quality large nuts.
> Fernor: Good cropper, large nuts, blight tolerant.
> Mars: A Czech variety. Good crop, but thin shelled.
> Jupiter: A Czech variety. An early good crop, but thin shelled.
> Saturn: A Czech variety. Large, oval and sweet nuts.
> Franquette: Not a heavy cropper with only one or two clusters.
> Buccaneer: Upright habit with a round nut.
> Broadview: Earliest cropper with small round fruit after three to four years.
> Rita: From the Carpathian Mountains. Produces a heavy crop with a thin
> shell. Compact and ideal for a garden.

It is obvious that the small garden is not suitable for large trees albeit the walnut tree has a long life and will grow large only after several years. A small garden needs small plants and small to medium size trees.

Without room, the branches will cross and in the end the tree will look a tangle. Another place where planting is not ideal is where there is a high hedge or building. The branches at the back will not grow out and the tree will take on a list or will bend forward. But some people will accept this disadvantage, just to have the tree and the nuts in the garden on its own and naturally grown.

Trees such as walnut may not fit into a garden where there is a desire to have any form of symmetry. Poplars or cypresses would be more appropriate for this kind of design. They do fit into avenues and with plenty of room show off well amongst other trees. There are some examples of trees in gardens and other locations in the photographs shown below.

I am impressed by Norwich City council which in 2017 sadly lost 15 red chestnut trees in Eaton Park. The horse chestnut has a lifespan of about 60 years and the city had to take the decision to fell them. The council and its advisors decided they wanted a collection of trees which would have longevity. They chose the walnut which they hope will last for 250 years.

A motorist's view of a walnut tree on the corner of a crossing in Bath. The tree has a fine shape despite buses and other large vehicles brushing against it. The Romans who brought walnuts to the United Kingdom would have been proud that one of their imported trees was planted in a city they founded.

Another view of a walnut tree found in the beautiful city of Bath. It is set among now-common features such as the litter bin in the foreground. The tree does not seem to mind its confinement under paving.

A charming smallish garden with a walnut tree at Dorchester-on-Thames, Oxfordshire. The tree did not seem to be likely to dominate the garden for a few years yet.

There is a good example of a 250-year-old *Juglans regia* at the Royal Horticultural Society's garden at Rosemore, Devon. The tree demonstrates how the wide-spreading branches provide excellent shade in a garden. It shows itself off well standing on its own. Yet the age of the tree and the weight of the branches has had to be considered to prevent damage. The RHS has cut some limbs to reduce weight. The tree itself is leaning slightly which I assume is caused by the wind.

Juglans regia at the RHS garden Rosemore, Devon, in early spring.

These outsized stone walnuts are beneath the common walnut at Rosemore. Quite a problem to crack these nuts!

Summer lunch or afternoon tea under the walnut tree, shading the participants from a hot sun, is a very European scene perhaps typical in the Dordogne in south-west France. In England the same scene might occur under an oak, beech or willow, but there is no reason why it should not be under the walnut. At the annual Glyndebourne Opera Festival, a picnic in the garden if the weather is right, is an essential part of the occasion. Helpfully the garden has two or three good examples of the common walnut which have been pruned in such a way that the opera goer can have an idyllic summer picnic shaded from the evening sun. But a word of warning: in the evening, birds come in to roost and the opera goers in the photograph below seem to have recognised the risk of sitting right under the tree.

Opera-goers enjoying a picnic beneath the walnut trees at Glyndebourne.

Philip Waites, the head gardener at Wimpole Hall says that the dappled shade and light through the branches is one of the great advantages of this tree. Although not much use to the opera goer at Glyndebourne, there are nuts on the Wimpole Hall trees which give the impression that the garden is productive as well as attractive. In order to get to a comfortable height, the trunk has to be kept clean as the tree grows. Preventing lower branches from growing out of the trunk can be done by cleaning or by picking them out. For larger branches, a quick snip of the branch with secateurs at a height of at least two metres can help. Then, there is a wait of perhaps

A comfortable seat at a small *masseria* (farmhouse) hotel on the Salento peninsula in Italy near Otranto, well shaded by a walnut at certain times of the day.

15 years. We pruned to this height because our motive was to get machines under the trees to keep the ground clean and pick up the nuts efficiently.

Pruning should be done in midsummer after the green walnuts have been taken. I look to train the tree back to a tulip shape. That means cutting branches that cross or hang too low. Then it is important to clear away cuttings to avoid the chance of disease.

There are however examples where the low branches have been left which has the effect of shaping the tree like a dome. There is a fine Chinese walnut at the Kew Gardens site at Wakehurst where the large bottom branches almost touch the ground.

The look of the tree from a distance is pleasing because of its shape. It looks more like a large shrub than a tree.

The tree is hardy in our climate. It is not grown in plantations here, but is a useful ornamental tree. It can grow to 25 metres. The interesting feature is that the leaf can be up to a metre long with 11 to 19 leaflets.

The small and extraordinary trunk of the Chinese walnut at Wakehurst. The tree is on the edge of a meadow with plenty of space around it. This tree is a relative of *Juglans regia*. The tree produces nuts, but a feature of the nuts is that they have very thick shells. The tree is a native of China and Taiwan and was brought to the United Kingdom in 1903 by Ernest H. Wilson.

The Chinese *Juglans regia* growing at Wakehurst. It is well displayed in this meadow situation.

A springtime photograph of a Chinese walnut found in Jersey. The flower at the end of the branch is a pronounced red colour. Attractive, but the catkins to the left, up the branch, are not mature. It is likely that the pollen from the catkin is mature after the stamens on the flower are fertile. Sadly, an annual event I am told, which is probably why the tree has never produced any nuts.

There are specimen walnut trees in some of our arboretums and in private gardens which have been imported by previous and current curators together with the owners of these sites. The *Juglans ailantifolia* is also known as the Japanese walnut or Heartnut. The tree has great amenity value and produces shade in the summer from its long leaves which can be a metre long. It first came to the United Kingdom in 1860 and grows to a medium size. The advice offered by sellers is that it should not be grown in areas known to be frost pockets. The nut is sweeter than other walnut trees and does not have a bitter after taste. It is characterized by its exceptionally long leaves.

There are a number of other specimen walnuts which will grow in the United Kingdom according to W.J. Bean's, *Trees and Shrubs Hardy in the British Isles*.

Juglans cathayensis is a 20-metre tree from China. The leaves are a metre to a metre and a half long. The tree was brought from central west China in 1903. It has a thick shell, but a small nut.

Juglans cordiformis is a 15-metre high tree from Japan. It is not common in the wild.

Juglans mandshurica is a 15- to 20-metre high tree, native of Manchuria in northern China. It was taken to St Petersburg by Carl Maximowicz.

Juglans ailantifolia at Westonbirt. The long leaves are very noticeable.

Juglans ailantifolia at Westonbirt. The distinctive leaf structure is clear from this picture.

The bark of *Juglans ailantifolia* at Thorp Perrow. This is a champion tree and the number is the catalogue number.

Juglans sieboldiana is a 15-metre tree from Japan with half metre long leaves. It was brought to Europe by a Mister Siebold in 1860.

The arboretum at Thorp Perrow has a collection of walnuts with some champion trees. Some of the arboretum dates back to the 16th and 17th century. These trees are the old parkland trees. A pinetum was planted in the 1840s and 1850s by Lady Augusta Milbank, but in 1931 and onwards Sir Leonard Ropner planted 60 acres of arboretum. A collection of walnuts was included. They have survived well in spite of the sometimes wet and cold climate and alkaline soil. I was told that trees surrounding the walnut collection seem to protect them from frost. The design of a collection is not the same as for a garden. Collections, as for instance at Kew, place the trees closer together for reasons of space.

Another reason for the walnut to be planted in a garden is its nuts. The walnut is prized for its part in Christmas lunch and for by those who enjoy pickling. I do not believe that the walnut is a fashionable choice for the gardener, landscaper or designer in Britain. It is thought of as a continental tree. It does not grow well in the northern part of the United Kingdom and wetter western parts of the country, and there are not that many nurseries where it is possible to acquire saplings. There are people who have walnut groves in Yorkshire (Thorp Perrow has a national collection), and you can see individual trees in Scotland, but late frosts are the risk to successful cropping. Equally, many of the people I have spoken to do not realise the timber value of the tree. I have heard of several gardeners with individual trees who describe the years when they see no nuts because of the thieving squirrels. It is obvious that a garden will not be so well protected against this menace as a grove from which the owner is reliant on income.

I have seen *Juglans regia* walnuts in some strange places. Whilst travelling down to the West Country, the railway leaves Paddington and the first stop is normally

Reading. As the train slows into the station and stops, the observant passenger will notice a few walnut trees between the tracks. How on earth did they get there? How long have they been there and how did they grow through the stones supporting the track? I wonder if there are nuts on the trees and whether the drivers put an arm out to collect them as they go by.

For a larger garden, I believe that it is beneficial to place walnut trees in clumps. A group of three or five planted 20 metres apart can look attractive. Walnuts are grown in this way in arboretums. There are fine examples in Kew, Wakehurst and Westonbirt near Tetbury.

Juglans regia elsewhere

The use of walnut for shade can be seen along the roads of Romania. Apart from the nuts and the eventual timber value from the trees after 40 years, the traveller can enjoy a pleasant walk or drive protected from the hot summer sun. In practical terms it is sad that the size of heavy goods vehicles has grown and grown. The endless brushing of the branches by these vehicles will reshape the tree and the shade will

This is one of two walnut trees that thrive near one of the oldest churches in France at Germigny-des-Prés near the banks of the Loire not far from Orléans. The church was built between 803 and 806 AD by the Bishop of Orléans, Thesduff. The trees could be 20 years old and were full of nuts in mid August.

An avenue of walnuts just below a terrace wall at Villa Della Porta Bozzollo not far from Lake Como. To the left is a row of roses. The terrace below had an avenue of cherry trees. Both avenues have been fairly recently planted, possibly just less than ten years. The trees seem to be leaning away from the wall.

The same avenue from above with a row of irises to the right, extending the whole length of the avenue. It is a very attractive design.

not extend so far into the road. Walnuts in Romania are grown wide apart enough to allow a substantial flock of sheep to lie down in the shade during the heat of the day. It is an ideal growing place for *Juglans regia.*

John Evelyn describes, in his book *Sylva Vol. I*, the Bergstrasse, which is a road between Heidelberg and Darmstadt. It is planted with walnuts for their ornament and shade which enables a man to ride for many miles under a continued arbour or closed walk.

Napoleon saw the benefits of planting along the roads for shade to protect his army on the march. He did not particularly use walnut for this purpose, but the Romanians did, seeing the benefit of the trees for nuts, for food and trade.

Juglans regia walnuts are grown in harsh climates. Canada can experience temperatures of minus 30 degrees in the winter and it would seem unlikely that walnut trees would survive in this kind of environment. Professor Crath was born in Kiev in 1893. He was a Polish agriculturist who migrated to Canada in 1908. He trained and became a pastor in the Ukrainian presbyterian church between the years of 1924 and 1936. He travelled back to the Carpathian Mountains in Poland and collected 20,000 walnut seeds which he brought to Ontario. As an interesting aside, Professor Crath would have known that the Poles refer the common walnut as the 'Italian' walnut. Had he been Russian, he would have referred to it as the 'Greek' walnut. He achieved good survival rates from the seeds despite the harsh climate. It is known that for the older tree, the risk is a late frost. Worse still could be frost of minus three degrees which has been experienced as late as May. The low temperature causes a lack of germination and the leaves to shrivel and die back. According to the Ontario Walnut Growers Association, Canadian people are more likely to plant walnuts in their back yards rather than in commercial groves. In France, and in particular in the Dordogne, you see pockets of trees and individual trees in gardens and elsewhere away from farmed land.

New Zealand has a different type of climate problem; the problem of wind. It seems that the bulk of walnut plantations are on the east side of the south island. High winds indicate that shelter belts around the walnuts are desirable. These belts will be planted with species which will keep the wind out even though some wind is required to allow the tree to germinate. The other issue is that New Zealand is much cooler than the northern hemisphere. The timber is exported as valuable gunstocks which compensates for fewer nuts.

Italy has good nut-growing areas. Whilst on a trip to Lake Como I noticed there were trees along the roadside. I also found some in gardens (see opposite).

6  The black walnut

Juglans nigra is the black walnut which is indigenous to the USA. It is a native of Eastern America, from Quebec to Florida, the Mid-West and the Great Plains. Frederick Stors Baker says, "The rolling country of south eastern New York, eastern Pennsylvania and New Jersey west of the sandy parts, where the soils are deep, present favourable sites for walnut. Western Maryland and the Shenandoah valley of Virginia are generally well suited but eastward to the coast the soils are usually not adapted, although occasional sites on river bottoms or rich flats are excellent and have produced walnut shade trees of large size. In Northern Carolina, Southern Carolina and northern Georgia except in the mountain region, it is doubtful if the

This splendid autumn leaf colour is on a black walnut grown by the author from seed.

The rounded nuts and the leaves of the black walnut.

climatic conditions are best for walnut." The *Juglans nigra* has attractive bark which is brown or black. The fast-growing tree develops a thick bark which is broken by cracks. The catkins are produced in the same way as the common walnut. The leaves are long and pinnated and the tree thrives in well-drained and fertile soil.

There is the *Juglans nigra laciniata*. The word *laciniata* refers to the leaf and the way in which it is cut. It is interesting that this black walnut tree will not grow in France. Yet its close relation, the *Juglans regia*, grows in abundance.

A variety of black walnut is the little walnut known as *Juglans microcarpa*. It is found along streams in Texas. It grows three to nine metres tall.

Although native to the United States, we British have known of this tree for a long time. John Evelyn in his book *Sylva* says, "The black [walnut] bears the worst nut, but the timber is much to be preferred and we might propagate more of them out of Virginia where they abound and have a square nut, of all other the most beautiful and best worth planting. Indeed had we store of these we should dispose the rest, yet those of Grenoble (*Juglans regia*) come next in place and are much prized by our cabinet makers."

One of the main differences between the common walnut and the black walnut is the catkin. The common walnut catkin is firm. The black walnut catkin droops.

Andrea Wulf writes about the black walnut in her book *The Brother Gardeners*. She says although the tree had been cultivated in Britain for some years, according to the botanist Miller, the black walnut was considered a rarity in 1731 even though it is thought to have arrived in Europe in 1629. John Bartram was a botanist collector or what can be described as a plant agent in Britain for certain large landowners. Bartram, in true Hollywood cowboy fashion, would set out on horseback, riding the wild places of America, facing dangers of wild animals, Native Americans and the weather to find unusual plants to send to Britain. Importing American trees and plants was fashionable back at that time. It was an age when impressing others was important. Andrea Wulf mentions in her book that Lord Petre bought 1,000 walnuts for his 1,000 acre estate at Thorndon in Essex. Bartram had obviously seen the mature black walnut tree whilst on his expeditions. Lord Petre, I can only assume, had only seen samples of young walnuts from seed and possibly some drawings. I rather doubt that he had much idea what the tree would look like when fully mature, 50 to 100 years later. In that respect, Lord Petre would have bought his collection of plants and trees with a lot of trust without knowing what his estate would look like long after his death. He was, I believe, a risk taker who wanted to create something unique.

After Lord Petre's death the garden design changed. His son became enamoured with the landscaping of Capability Brown and changed the estate to Brown's plans.

A fine black walnut tree at Ballykilcaven, Ireland. The tree is in a good position as it has plenty of room and it shows itself well.

A black walnut amongst other trees and bracken along a path at Wimpole Hall.

The black walnut is a fast-growing tree which can grow up to 15 metres high with a wide head and a dark trunk if the tree is in the open. I have grown black walnuts from seed and in the first year they can grow half a metre. In a forest situation they will grow to 30 metres and the top branches or crown will become stag-head shaped as the tree gets older. The trees in a forest are crowded, if planted by man, so that the lower branches do not grow. The canopy becomes dense so that weeds are crowded out. The advice is that a thinning programme to manage the forest is necessary. Planning this thinning will be considered after about 25 years. The University of Missouri says the thinning may involve taking out every other row.

Like any project, economics must come into the picture. The obvious measurement is the number of trees planted per hectare. That is certainly the case with soft- and indeed many hardwood plantations. Bearing in mind the remarks made by the University of Missouri the planter will be left with wide rows. That indicates that a calculation needs to be done along the lines of the quantity of timber compared with value per hectare.

Black walnuts need a rich loam soil but can grow in poorer soils. The University of Missouri says that the pH should be six to seven. The better the fertility, the better the growth. What is most important to the growth of the black walnut is light. They do not tolerate shade, but they will grow in close proximity to other trees so long as they can compete.

This black walnut is said to have been planted by Henry Drummond in about 1850 in the garden of Albury.

The black walnut growing in the Evelyn garden at Albury was grown from a seed planted by Henry Drummond. Known as *Juglans nigra alburyencis*, it is also found in Kew gardens.

A beautifully-shaped black walnut at Albury.

Albury is in Surrey. It was owned by Thomas Howard who died in 1646 and it passed to Henry Howard who became the Duke of Norfolk. John Evelyn helped with the garden planting. On Henry's death the estate passed to Heneage Finch. He became Earl of Aylesbury and sold the estate to his brother Captain Finch. In 1784, the captain's son inherited the estate and through various other owners it came to Henry Drummond in 1819. Drummond altered the house with the help of Pugin and planted many exotic plants. There is a collection of black walnuts in the park.

The black walnut *laciniata* leaves at Wimpole Hall. The leaves are split which makes them look thinner, as if they have been cut by a knife.

A black walnut on its own in the park at Wimpole Hall.

This tree is in the Brera botanical gardens in Milan. It is a *Juglans microcarpa*. It is also called the little walnut and comes from Texas and Mexico. It likes to grow along streams and is a variety of the black walnut.

In a United States forest, nuts will be produced, which fall to the ground and this creates a chance of "natural regeneration". The tree, through the seed, is creating a sapling allowing itself to reproduce in its own genetic likeness. In reproducing itself, the tree might be of the finest quality or it could be a poor sample. Selection by man does not come into the growing of the sapling in the forest. For this natural regeneration to happen, the nut needs to be buried. A coating of leaves, providing it is thick enough, might do the trick.

For once the squirrel, or similar animal, seems to have a benefit. It will find and transport the nut away from the tree and if it buries it in an area where there is light, there is a chance that the sapling will grow. There is, however, another hazard the young sapling must face and that is the grazing deer. Deer are shy animals and the forest fulfils their desire to keep themselves hidden. In some areas of the walnut forests there is domestic animal grazing, in particular cattle and pigs. Grazing keeps the forest floor clean and there will be no natural regeneration.

The black walnut will live for more than 250 years, but the average is 90 years. The timber is of fine quality and has many uses which I will describe later.

There is a prolific growing area in Oregon where walnuts are "larger, finer, flavoured and more uniform in size than those grown elsewhere," according to Jacob Calvin Cooper who wrote *Walnut Growing in Oregon* in 1906. He went on to say that they are free of oiliness and contain a full meat that fills the shell. "The best areas to grow them is about 200 acres in Oregon". But according to Cooper, "the richest land known to man is the Willamette basin with its tributary valleys and hills. This area is 60 by 150 square miles." The walnuts themselves are slightly tapered at the bottom. It was claimed that the strains planted in 1908 in Oregon became properly fruit bearing after eight years, whereas the European walnut will only bear quantities of nuts after 16 to 24 years. It is a point of view that contrasts with John Evelyn who clearly did not like the taste of the black walnut. His dislike may have stemmed from the fact that the walnut he tasted could have travelled a long way and may have been badly stored since it was picked.

Black walnuts need deep clay loam soil, principally so that the water can be retained during the summer months. Oregon does have a possible problem with what Cooper calls "hardpan". The phenomenon is where the soil has become very compacted. The effect is that the roots are not able to perforate this layer. The hard layer can be broken up by heavy machinery. One suggestion in Cooper's book is to auger a hole a metre to a metre and a half deep. Then, in order to break the pan, throw in a lit stick of dynamite to loosen up the soil. The black walnuts have big taproots which need to able to spread and therefore the breaking of any pans is nothing but helpful. The rainfall is critical with this type of tree as it needs about a metre of it. It is planted in eastern Europe for timber but is said to grow poorly in France.

The tree will grow with few side branches if crowded by other trees, but in the open it will grow to a great width. In a forest where natural regeneration takes place the trees will grow close together. The lack of side branches is good for timber as there are no knots.

Pliny wrote about allelopathy in 77 AD. Allelopathy is the secretion of a biochemical material that prevents the germination and growth of vegetation under a tree. Eliminating competition under a tree whilst it is growing is a good thing. The black walnut (not the common walnut) secretes this chemical called hydrojuglone. The effectiveness of the hydrojuglone will depend on the soil conditions such as drainage, soil aeration, temperature and microbial activity. Soil microorganisms will take in this kind of chemical and take out its toxic effect. The better the soil the better the chance the under vegetation has of surviving. The urban soil is generally poor and plants susceptible to the chemical are more likely to die.

In Britain I have not seen much sign of this phenomenon. I believe this is important because the British amenity planter might not plant the tree if there was bare earth under the tree. In a wooded garden, I like to see some plant layering even up to the trunk.

The work of Franciscan monks in California

Franciscan monks are said to have brought the black walnut tree to California in the early 1800s. Few black walnut trees in California are a pure strain. They are American and eastern black crosses. Many are black walnut rootstock, quite often *Hindsii*, with grafted English or French scions. Even in 1908 strains such as *Mayette* and *Franquette* are the first choices. Others include *Praeparturiens* which has a fine flavour like a Hickory nut. *Charberte* is a hardy tree, which is good in upland areas. The *Ford Mammoth* and *Gladsy Byon* are considered too large.

The black walnut is recorded as having come to Great Britain in the 17th century. King James I of England, issued charters for the colonization of what was later to be known as the United States. As a consequence, Captain Christopher Newport set off for Virginia. He was followed by Captain John Smith who became a community leader and who helped found the new settlement at Jamestown. He was a participant in the romantic story of Pocahontas who is famously said to have saved Smith from execution by her father by standing over him. Later, she was taken hostage and taken to Jamestown where she married local magistrate and tobacco grower, John Rolfe. She came to London, intriguing its inhabitants who had never seen a Native American woman before, but sadly she died en-route down the Thames whilst returning to Virginia. She is buried in Gravesend. The story does not have a direct link to the arrival of walnuts from America because, when the early settlers arrived they sent clapboard back for building. The connection is that she came from a part of America where the black walnut grows naturally. She would have walked in the forests and her male relatives would have hunted through them.

Colonization was not easy in Jamestown Virginia and the new colony nearly did not survive as the population suffered disease, native attacks and the weather. It is recorded that black walnuts, native to that part of America, were exported from the colony to the UK beginning in 1640. The main imports began in 1729. Walnut timber did not form the backbone of the new colony's wealth into the future, only later did it become apparent how valuable the timber was for furniture making. It was tobacco that became the most important commodity for export.

Looking up at a black walnut at Kew.

Saplings were brought to the United Kingdom, but many did not survive and some were thrown overboard by pirates as unwanted booty. There are examples of these beautiful trees in Battersea Park, Kew Gardens and Wakehurst Place.

At Wakehurst Place there are mature trees in the garden which stand on their own. There is also a young tree area at Wakehurst known as Horsebridge Wood that

A close up of a black walnut trunk at Kew Gardens. It is the rough bark that is characteristic of the tree.

A young black walnut in Horsebridge Wood, Wakehurst Place in early spring.

A black walnut at Westonbirt.

has been planted with an American theme. There is a mixture of American oaks, shag bark hickory and black walnut.

The hickory is also a native of North America and is a close relation of the walnut. Hickory provides a nut and, famously, the timber was used to make golf club shafts. The American trees, like other woodland areas at Wakehurst, are planted quite close together. There is a real benefit to this planning because it means that thinning can be done selecting only the very best examples.

Probably the oldest black walnut tree in England is in London at Marble Hill. It is thought to have been planted at the beginning of the 18th century. Marble Hill was

Marble Hill House, Twickenham. The home of Henrietta Howard, Countess of Suffolk.

The old black walnut is fenced off with no public access under its canopy.

The black walnut at Marble Hill.

built for Henrietta Howard, the Countess of Suffolk, mistress of George II. Building of the house was started in 1724 and completed in 1729 to create an arcadia, or peaceful haven away from the hustle and bustle of London life. Alexander Pope, the poet and a friend of the countess who lived nearby in Twickenham, may have had an influence on the creation of the garden. The upper and nearest lawns to the house were said to be for bowling and drifting down to terraces at the edge of the Thames. On either side of the lawn, there were horse chestnut glades.

Just outside the doors of the north transept of St Paul's cathedral in London there are four black walnuts by the gates heading out to Paternoster Square. There, on the left hand side is a statue of John Wesley shaded by two of the black walnuts. These trees are young, over time they will grow possibly shading out light from the nearby offices.

Two black walnuts shading and sheltering a statue of John Wesley near the north transept of St Paul's cathedral, London.

7  Trees in other parts of the world; diseases

There are other walnuts of the *Juglans* species. Included in this group are trees in Argentina, Arizona, Brazil, California, the West Indies, Texas, Nuevo Leon, Mexico and the *Hindsii*. The *Juglans jamaicensis* is a native of Cuba, Haiti, the Dominican Republic and Puerto Rico. It is an endangered species and is found in forested areas of the Monte Guilarte Commonwealth forest between 700 and 1,000 metre altitude. The soil is clay and it requires about two metres of annual rain as indicated by the US Fish and Wildlife Service Recovery Plan for the West Indian Walnut.

Juglans hindsii

According to Frank Callaghan, who in 2008 wrote a piece entitled *Hind's Walnut* (*Juglans hindsii*), this is one of only two species on the west coast of the United States in California. In 1837 Richard Brinsley Hinds was on an expedition, including plant

An example of *Juglans hindsii* at Wimpole Hall.

hunting, aboard the British ship HMS *Sulphur*. They landed and marched up the Rio Sacramento from San Francisco. The expedition was recorded by George Bentham who wrote the *Botany of the Voyage of HMS Sulphur*. It took 70 years for the walnut tree to be named after Hinds. Yet, remarkably, the collection created by Richard Hinds is at Kew. There are other examples in National Walnut collections such as at Wimpole Hall.

The largest specimen is currently found in the Napa Valley. There are trees in the north of California and also in Oregon. The tree has a rapid growth and can live to an age of 300 years. The wood from the *hindsii* is well known for the production of high-quality gunstocks as well as in furniture making. The largest tree recorded was twenty-nine feet (8.8 metres) in circumference! Callaghan says that the *hindsii* walnut grows best in a Mediterranean climate with a mixture of warm summers and moist winters.

The other Californian Walnut is the *Juglans californica*. It was found, according to Frank Callaghan, in the Sierra Santa Monica, 23 years after Hind found the walnut finally named after him. It thrives in the southern part of California. The warmer climate is essential for this species as it is susceptible to frost. According to W.R. Bean it will grow in the United Kingdom, but is not very common. This species is smaller than the *Juglans hindsii*. It is five to ten metres high and has several stems coming from the ground, giving it a sort of coppiced look. In America, the rootstock is used for a *Juglans regia* graft to create a crop. As both the *Juglans* and the *Californica* are native to the USA, it is a natural fit.

Juglans cinerea is the butternut. It grows up to 30 metres and has a paler bark than the black walnut. The leaves are 25 to 50 centimetres long. The male catkins are five to ten centimetres long. They were first brought here from the eastern USA in the 17th century and a good sample can be found in the Westonbirt arboretum. The Japanese walnut is similar to the butternut. Its leaves are 90 centimetres long and the nuts are round.

Walnut diseases and pests

Walnut blight otherwise known as *Xanthomonas campestris pv. Juglandis* is a debilitating disease. It is a bacterium which spends its winters in buds and catkins, according to studies published in an online guide to plant disease control issued by Oregon State University Extension. When the spring arrives the bacteria spread along the growing shoots and nuts.

There is a critical period when, if there is prolonged rain, there could be a severe outbreak. This period is two weeks before the bloom and can have a severe effect on the nuts.

The first sign of infection is reddish brown spots on the stems. Young infected leaves and catkins turn dark brown or black and then die. The nuts, when they arrive, are also affected by black slimy spots. The organism will penetrate the husk, the shell of the nut and occasionally the nut itself.

It is a hideous disease on a majestic tree. What can be done about it? The solution, as suggested by the University for Californian Growers, is to spray at the "pre-bloom" stage. It would seem that the spray adjuvants are copper based but there are now some bacteria which are resistant to copper. The copper solution has been used for 40 years. Some growers have tried to spray after rain but this has not worked very well. In order to gain some success, Messrs Olson, Buchner, Adaskaveg, and Lindow say that up to ten applications of adjuvants should be applied. These applications should be applied every seven days. It was recognised in 1990, that some strains of the bacteria have become copper resistant. That said it would seem that the copper resistant strains are less damaging to the nuts. The guide does however say that silicone-based adjuvants show promise.

I have seen this in my own trees and it is a sorry sight. With the information that the blight is weather-, and in particular rainfall-dependent, it is not surprising that we have blight in the United Kingdom.

The biggest pest is the codling moth. This moth can damage the fruit. It came from Europe and is now found in England. Having got into the nut the larvae will cause the nut to fall. The female moth will lay eggs in the bark of the tree. The larvae come to life in March at the time the leaf and bud are emerging. A female moth could lay up to 30 eggs. The solution is said to be spraying or pheromone treatment, which works as a mating inhibitor.

Another equally devastating disease is honey fungus. This disease spreads underground and prevents the roots from supplying water to the trunk and branches. The upper part of the tree dies and the bark cracks. The roots below become rotten. The botanical gardens in Cambridge had a black walnut tree said to be 150 years old. Honey fungus got into its roots and it had to be cut down in 2017.

This list does not include all the diseases and pests which could damage the walnut.

8  The timber

This very charming poem written by the American poet Mary Oliver explains the dilemma to fell a tree and pay off a mortgage or to just enjoy its beauty.

The Black Walnut Tree
My mother and I debate:
we could sell
the black walnut tree
to the lumberman,
and pay off the mortgage.
Likely some storm anyway
will churn down its dark boughs,
smashing the house. We talk
slowly, two women trying
in a difficult time to be wise.
Roots in the cellar drains,
I say, and she replies
that the leaves are getting heavier
every year, and the fruit
harder to gather away.
But something brighter than money
moves in our blood – an edge
sharp and quick as a trowel
that wants us to dig and sow.
So we talk, but we don't do
anything. That night I dream
of my father's out of Bohemia
filling the blue fields
of fresh and generous Ohio
with leaves and vines and orchards.

What my mother and I both know
is that we'd crawl with shame
in the emptiness we'd made
in our own and our father's backyard.
So the black walnut tree
swings through another year
of sun and leaping winds,
of leaves and bounding fruit,
and, month after month, the whip-
crack of the mortgage.

Timber has always been an important commodity for human beings. Its use in building has been important since mankind needed shelter and warmth.

How is timber created?

Photosynthesis is the conversion of sunlight into sugar which is essential for the growth of any plant or tree. Carbon dioxide is taken up by the leaves and water through the roots. Oxygen is released through the leaf into the air. Maricopa Education say that the chemical equation can simply be described as six molecules of water plus six molecules of carbon dioxide produces one molecule of sugar plus six molecules of oxygen.

The growth is one annual ring or layer after another inside what is called the cambium layer. The inner part of the tree is called the heartwood and is hard. The outside is the sapwood and is the highway for the sap to get to the crown. The heartwood is darker in the walnut tree. Curiously, although the heartwood is enclosed by the sapwood and bark, it is vulnerable to disease. Yet, when the tree is felled the hardwood becomes harder and resistant to decay and rot.

According to an old farmers' almanac, the black walnut timber was used for fencing and building materials because of its resistance to rot.

The timber for furniture needs to be clean. That is to say without knots. The walnut is a tree that has a productive life producing nuts and may have an even longer life once it is felled. The tree will produce nuts for many years and can then be considered for its other use – timber. The pruning of the tree in the first years after it is planted is vital not only for the nut production but also for the timber. The main trunk height may be governed by the necessity to pick the nuts in a practical way. The required height from ground to the first branch might be two metres. In order to get the best timber, those two metres of trunk need to be without any branches, that is to say clean. The quicker the side branches are removed, the less damage will be done to the trunk. It is the same with any timber tree. Oak for example used to be grown with a nurse crop. This was hazel which was coppiced every two to three years so that it pushed up the oak and shaded out the lower branches. Woods with this combination can still be seen today, but young hazel coppices are no longer used to make sheep hurdles and therefore the hazel is not used. Today, young trees are protected by a plastic tube and this has the same effect up to the metre length of the tube. Pruning as the tree bushes out of the top of the tube, pushes the tree up and prevents it from bushing sideways. Walnuts have pinnate leaves and they do not thrive in a tube. The problem is that the tube confines the walnut leaves which want to spread.

The colour of the walnut wood varies from light to very dark. European walnut is moderately hard, heavy, strong, and tough with a medium texture. Black walnut is uniformly darker, heavier and more resistant to furniture beetle. A buyer or dealer of timber for furniture must judge what is inside the log from the outside, which requires skill. He can inspect the tree as it stands or view the end of the log when it is felled. He must imagine the colour and graining inside. Only when the log is cut open on the saw bench will the dealer really know what he is selling and how useful it will be for furniture makers. Apart from the colour, which is the most important feature, there is the patterning or grain to consider. The furniture maker has a choice. He will either use the best veneer he can find, which he can see and judge for its colour and graining when he selects it, or he will use planks or whole timber.

Up to 1720 France was the main exporter of walnut to Britain, but this stopped because of an export ban. This was not unusual as Britain and France were constantly quarrelling or at war during the 17th and 18th centuries. Adam Bowett in his book *English Furniture 1660–1714 from Charles II to Queen Anne*, points out, "Furniture was surprisingly vulnerable to these commercial pressures because of their dependence on imported raw material. The importation of French walnut was stopped when war broke out in 1689".

To make matters worse, tariffs on furniture were imposed to help pay for the war. That did not stop a continuing demand in England for furniture and imports built up from Holland, Spain, Germany, Italy and Turkey. Yet, the master furniture makers to the Crown, Jensen and Roberts, claimed that the French Grenoble walnut was best.

In 1720 the import duty on timber was removed. Most of the imported walnut from the United States came from Virginia and Maryland. The Virginia walnut was particularly favoured by the chair makers. Furniture making near the ports of Bristol, Liverpool, and Lancaster grew and the boom was to last for ten years. After 1730 the popularity of walnut as a timber was overtaken by mahogany. At the same time there was a devastating period of disease in the walnut population of France.

Walnut planks
Walnut planks from the trunk can be used for plain furniture, kitchen worktops or even flooring. The kitchen worktop needs a lot of oiling to protect it against wetness,

heat and other damage. The flooring needs to be well matched up to ensure that the graining is constant.

In the Dordogne region, the trees are kept in nut production for about 15 years. I was fairly surprised to learn from one producer we spoke to that he had no interest in the timber. I think that this meant he did not get much money for the redundant trees once their usefulness as nut producers came to an end. Yet some of the timber is used for local furniture making. There is a fine walnut fire surround at the Château of Hautefort. Burr walnut is prized for its natural design and tight grain. It is cut from diseased branches. Some fine walnut chests or chests of drawers were made in Georgian times.

There are some fairly obscure reported uses of the timber. The hubs of early air-craft propellers like the Sopwith Camel were made from walnut. The reason being that the wood is easily carved, hard and strong.

The doors at Saint-Sauveur Cathedral in Aix-en-Provence

In Aix-en-Provence there is a surprising walnut treasure. It is to be found in the old town; in the Place de University is the cathedral of Saint-Sauveur. The cathedral itself is a combination of three churches, built side by side. As you look at the main door from the square you see some weather beaten mauve doors, which are situated at the entrance to the 11th-century church. The sun seems to have faded them badly. Yet these are a façade to protect the real doors behind them. Such protection was needed against the weather and the destructive intent of the French revolutionaries who destroyed many church treasures.

Inside the protective encasements is a real masterpiece of walnut carving. I am not a historian, but this took my breath away. The carvings themselves are a com-bination of statues and ornamental foliage. The doors were commissioned in 1505 on the orders of the canons who formed the Chapter at the cathedral at the time. Two carpenters from Aivois were employed but the first attempt failed for unknown reasons and the canons hired two local brothers Raymond and Jean Bolhit, and a sculptor called Jean Guiramand who started work in 1508.

Guiramand's doors are carved from solid blocks of walnut. From my observation these are straight half-round timbers which are clean with only a small number of small knots.

The carvings and statues are interesting. Decorations above the figures show symbols of the Eucharist such as acorns, grapes and pomegranates. The four prophets of the Old Testament, Isaiah, Ezekiel, Daniel and Jeremiah are featured. Other carv-ings feature "sibyls", sibyls were prophets or soothsayers.

These sooth sayers were important for propaganda because unlike the modern weather forecasters and analysts we hear today, they were supposedly capable of foreseeing the future. Their origins were in Greece, particularly Delphi, but these doors depict sybils predicting the birth of Christ, elements of his life, death and res-urrection. The Pagan sybils accompany other prophets from the Old Testament, thus creating a magnificent set of doors with a Christian message.

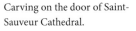

Carving on the door of Saint-Sauveur Cathedral.

Details of carvings on the doors of Saint-Sauveur Cathedral.

Carvings on the doors of Saint-Sauveur Cathedral.

In Uzbekistan and in other countries along the old Silk Road which used to be behind the Iron Curtain in the southern Soviet Union, doors are often made of walnut. As walnut is indigenous to this part of the world it is not surprising that the local builders use this timber for its strength and resilience to weather.

Saint Catherine of Alexandria's Basilica in Galatina, Puglia, Italy

Saint Catherine is famous for being martyred on a wheel. There is a relic, her finger, which is kept in the Basilica at Galatina.

The Basilica is not far from Gallipoli in Puglia on the Salento peninsula in Italy and is a gem. It is part of a monastery inhabited by silent Franciscan monks. It was first built in 1391 with money from Raimondello Orsini del Balzo. The arrival of the monks was on the orders of the Pope who commanded them to teach the Latin rule in order to counter the spread of Greek rule. The monastery is unique because of the frescoes ordered by Raimondello, his widow, Maria d'Enghien, and his son, Giovanni Antonio Orsini del Balzo. The frescoes were the work of the friars.

Churches and basilicas need pews and in the museum there are some pews made of walnut. They are well carved which can be seen in the photograph opposite.

A fine walnut
church pew in
which to attend
a service, but not
very comfortable.

The Scotsman building in Edinburgh

The Scotsman building was the headquarters of the famous Scottish newspaper. It was opened in 1903. The building stands above Waverley station near to the beginning of the Royal Mile in Edinburgh. The reception to the newspaper office was on the fourth floor at street level with the road to the bridge. The public used the door into the reception, but around the right hand side of the building is another entrance. This door was reserved for the directors of the newspaper. Once inside this door one finds oneself in a walnut-panelled room used to entertain the directors' important guests. The newspaper has since moved out and the building has been converted into a smart hotel. But happily the architects have kept many of the original details in the building. The main reception is now an oak-panelled brasserie and the directors entertaining room is the hotel's main reception. *The Scotsman* librarian described this room as the Walnut Room.

Carvings on the panels at *The Scotsman* building in Edinburgh.

The architects were Dunn and Findlay. The panelling came from Italy and was installed by Whytock Reid and Co.

Hever Castle

Hever Castle in Kent was home to Henry VIII's second wife, Anne Boleyn. Records of the castle go back to about 1200 and it has a rich history. William Waldorf Astor,

The Inner Hall at Hever Castle, Kent.

the first Viscount Astor of Hever, bought the castle in 1903. The panelling and pillars in the inner hall which was formerly a kitchen, are made of Italian walnut and are of exceptional quality. W.S. Frith carried out the work in 1905 as part of a restoration project. The colour and graining have been carefully selected; a golden colour with well defined black ring lines. The pillars, carved with fruits, seem to have been created from single trees. To the right of the hall is a staircase leading up to a gallery, the balustrade of which is carved in a similar manner to that found at King's College, Cambridge. The images on these fine carvings are roundels of human forms with Tudor roses spaced in the design. Obviously little expense was spared on the project.

The staircase at Dyrham Park

Dyrham Park is a late 17th-century English house just off the M4 on the road to Bath. Set in the south Gloucestershire countryside with 250 acres surrounding it, it is now owned by the National Trust. It was built on the instructions of William Blathwayt (1649–1717) who was Secretary at War under James II and is home to a large collection of Dutch Master paintings. Blathwayt was well connected in the society of his day; his uncle was Thomas Povey, the powerful one-time Minister of Plantations under Oliver Cromwell. Povey later became Minister of Finance to James, Duke of York. Well travelled and well read, Blathwayt was very fussy about how he wanted his house to be built, clearly he had good taste and an eye for craftmanship. The first design of the staircase was rejected as he thought there could be dust traps in the structure. The staircase that was ultimately built rises from the ground floor to the

The staircase at Dyrham Park near Bath.

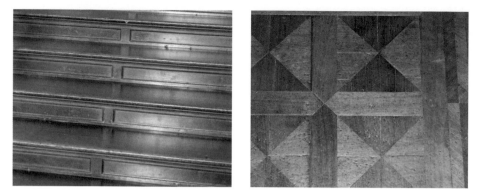

The stairs and parquet walnut floor on the landing of the stairwell at Dyrham Park.

first floor and is in a square formation. It is built of black walnut which was brought over from Virginia to nearby Bristol and constructed by Robert Barker of London.

Time passed and the family rented the house in 1938 to Lady Islington, the widowed wife of the first Lord Islington who was Governor General of New Zealand. During the Second World War she converted the house as a nursery for West Ham (London) children. She did extensive redecorating to the house which included painting the staircase bannisters. She probably disliked the darkness of the wood and painted the stairwell a grey colour. She left the stairs themselves, which could have been carpeted, but this has left a well patterned French parquet half-landing and the original boarded stairs.

Furniture

The human race needs furniture. We have learnt about comfort but we also appreciate items which are elegant, artistic, practical and provide storage. Tastes change over centuries, technology and materials change and peoples' needs alter. The human race began with a space on the floor to sleep, then we found it more comfortable to sleep on straw and eventually we designed the bed as we understand it today. This demonstrates how our designs have evolved throughout history to suit us, just as we decided sitting on a chair was preferable to sitting on the ground. The Tudors used oak for their furniture, but at the time of the Stuarts, the furniture makers discovered walnut to be fine timber and it gave them the opportunity to create new designs. The walnut itself is strong and can easily be carved and shaped. Added to this, it is a lighter wood than oak and has distinctive patterning in the grain. The first walnut timber came in from Europe to the port of London. Many of the joiners and cabinetmakers were established in areas that became the financial part of the city such as Aldgate, Holborn, Fleet and Houndsditch.

In 1721 the government passed the Naval Stores Act. The effect of this was to lift import duty on timber from the Americas. Walnut was imported, but there was a change in taste because the furniture makers began to use mahogany. Ships having crossed the Atlantic would return with mahogany planks if they were not carrying cargos of sugar or tobacco.

The Worshipful Company of Joiners, ranked 41st in the modern Livery company hierarchy, influenced the manufacturing of walnut furniture through their apprenticeship schemes. The makers of walnut and cane chairs also used the skills of the Worshipful Company of Basket Makers.

The skills of these craftsmen did not go unnoticed. Amongst many, the furniture maker Richard Robert made walnut furniture for George I and George II. Quality furniture was sent to Tsar Peter I in St Petersburg. Many of the great houses of England were furnished with walnut furniture. These houses included Montacute, Blenheim, Beningbrough and Althorp.

The Eglantine or "Aeglantyne" table in the High Great Chamber at Hardwick Hall, Derbyshire. The walnut table top is elaborately inlaid with marquetry depicting musical instruments, games and heraldic references. The table was made to celebrate the triple marriage contract between the families of Talbot and Cavendish in 1567.

A finely carved walnut chair made between 1560 and 1590.

A walnut chair with a woven seat.

Made between 1725 and 1740 in London, this chair features a carved seat made of beech with a walnut veneer and grey club rush. The rush has been replaced, but the rest is original.

A single George I walnut chair with a shaped back, cabriole front legs with Acanthus carving on the knees and arms presumably added at a later date, terminating in carved masks. There is a drop-in seat covered with red brocade needlepoint.

The attraction of walnut is its grain and the craftsman's ability to carve it into shapes. Carving is a particularly useful way to improve and enhance the look of chairs or tables. I show some good examples of carved legs in photographs in this book. I am particularly drawn to the Venetian chair, not because of its age, but by the way it has been so skilfully carved. The Eglantine table is a fine example of carving skill with finely turned legs and exceptional marquetry depicting musical instruments and games. One can only say what a generous present this was to the Talbot Cavendish young marrieds.

Furniture at the Royal Château of Blois.

By pure luck, I came across a fine collection of French Furniture in the Château of Blois, situated in the Loire valley. Seven kings and ten queens of France lived in the château. A period of restoration was started in 1845 after the château was neglected following the French Revolution.

This fine piece of furniture in the Château of Blois was made in Spain in the 16th or 17th century. It is known as a bargueño and is made of walnut, ivory and wrought iron.

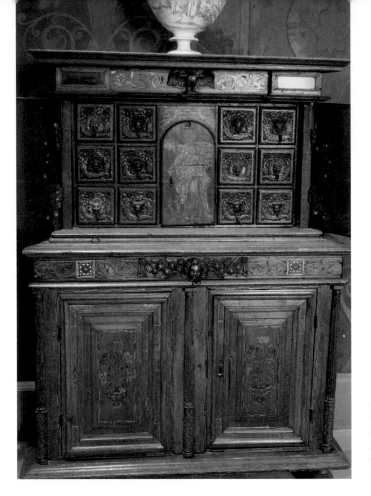

A two-piece walnut
cupboard built *c.*1600.
It is made of walnut
and coloured wood
with bone and marble
inlay. Château of Blois.

The big chair in the Milan Museum made in
the 18th century.

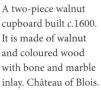

A 17th century Cassatone chest of drawers in the Milan Museum. The Bambocci decoration is characteristic of the Genoa area. Carved figures on the front and side may have been added in the 19th century.

A 17th century chest of drawers in the Milan Museum. Walnut inlaid with rosewood.

There are 564 rooms of which 100 are bedrooms and there are 75 staircases.

Built in the middle of the town that it effectively controls, the Château of Blois comprises of several buildings constructed from the 13th to the 17th century around the main courtyard. The "Salle des États Généraux", built at the beginning of the 13th century, is one of the oldest seignioral rooms preserved in France.

The furniture was built mainly of solid walnut.

Veneers

Veneers are thin slices of highly-patterned wood glued onto panels of soft wood to create high quality furniture or interiors.

Veneers are a fashion and it is a fact that fashion and popularity moves with time. The 1980s saw a revival in veneered furniture and this has remained the case since that time. The veneers themselves are decorative and people will buy products because they are quality items with colour and decoration.

Yet veneers should not be seen as the only way to create a fine piece of furniture, there are many furniture makers who make fine and valuable pieces from planks. Conversely, there are fine examples of sideboards, dining-room tables, coffee tables, card tables, pianos and many other furniture pieces are all made with a veneer exterior. Doors and panelling can also be veneered.

Bill Cleydert, who makes bespoke furniture, says he will always use veneer. He says he has been known to veneer over walnut itself.

Modern furniture makers use MDF as the base for the veneer. MDF (medium density fibreboard) combines hardwood and softwood fibres with resin and wax at high temperature and pressure to form stable boards. It is relatively cheap and has the advantage of being almost completely flat. This is essential for veneering not only for its presentation, but also to give it longevity. Any piece of furniture needs

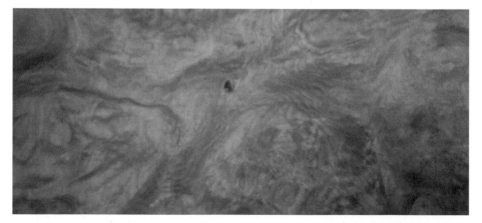

A burr walnut veneer. It appears grey and colourless at this stage. The colours are brought out later in the process.

This is a plain veneer where the knife has sliced straight down the grain. It appears grey because there is no lacquer on it.

A burr walnut veneer as part of a card table where lacquer has been applied.

to be made so as to withstand changes in temperature. Veneer on MDF will do this even with modern central heating. There are some climates where temperature is so extreme that there may need to be an aluminium backing. Car dashboards are made with aluminium frames, as we shall see later.

The veneer itself is bought from dealers from different parts of the world. The European walnut is slower growing than the American black walnut but the buyer is looking for a straight grain or one that is decorative such as the burr. That smoky-looking design is typical of this decoration.

The types of veneer are:

Rotary Peeling: The logs to be veneered are put on a lathe and revolved so that the veneer is peeled off.

A highly important George I burr walnut cabinet with arched and moulded broken cornice inset with two segmental verre eglomise panels decorated with sunflowers framing a lozenge-shaped mirror panel above two arched and bevelled mirror panelled doors (one panel replaced) enclosing a fitted interior with pigeon holes racks and drawers in various sizes. All surrounding a central concave niche inlaid with a sunburst and mounted with a gilt metal mask and scroll spandrels above two curved doors framed by Corinthian pilasters with gilt wood capitals each surmounted by standing figures of Mars and Minerva. The sloping base enclosing a serpentine fitted interior of the same quality with a central convex cupboard. Two doors enclose a mirrored interior with chequered and rayed door flanked by pigeonholes and small drawers and well below. The main base has two short and two long drawers cross-banded and inlaid throughout with fine chevron-pattern lines. Original gilt brass engraved lock plates, loop handles and three pairs of carrying handles. The whole on bracket feet.

Walnut-veneered bureau bookcase, George II period, *c*.1730–40. Comprising a desk with sloping top above four drawers, and a cupboard above with two doors. Walnut veneered on a pine and oak carcase. Top section with a cross-grained overhanging cornice made from a series of mouldings including an ogee, cove and frieze section. Below a pair of panelled doors with herringbone banded edge. Inset with dome-topped fielded panels, which are quarter veneered with herringbone banded edges, the frame with a cross-grained moulded edge. The frieze with two candle stand slides with small brass knobs. Interior – the doors with cross-banded edge and the same on the edge of the panel. The back of the cupboard grained to imitate walnut. Three shaped and moulded shelves, the bottom one with two pairs of bow-fronted drawers under and three square pigeon holes in the centre, the doors with their original hinges. An applied moulding with a cove and a bead, where the top sits on the base section. The fall with four veneered panels, with a herringbone border, cross-banding with a herringbone outer border and a moulded edge. Rests on two pull-out supports with brass knob handles. Interior – shaped and stepped with a centre door with a herringbone border with similar arched border in the centre, fitted with a drawer inside. A pair of straight fronted drawers either side and two either end. The base fitted with a well with a sliding top. The interior of the fall with a rectangular shape with rounded ends fitted with tooled black leather and with a cross-banded outer edge. Original hinges and knobs to the drawers. On the front the frieze which encloses the well, with a herringbone and banded edge. The two short and two long drawers below, divided with a double bead moulding. The small drawers with halved veneers and the large ones halved where each handle fitted, both with herringbone banding. Pigeon holes above. With later handles, originally they were held by wire fixings, the third set now.

A sewing machine used to join veneer panels.

Press used to mount veneer on MDF. The press can produce a pressure of 3,000 lbs per square inch at a temperature of 70°C.

Parallel slicing: Slices are cut from the length of the log. The effect is a variegated figuring. This is known as a crown.

Quarter slicing: The veneer is cut perpendicularly to the medullary ray, which radiates through the wood. It produces a comb effect.

Half round slicing: This produces a straight-grained figure by slicing from a quarter log parallel to the annual growth rings of the wood.

Rift cut: A segment of log mounted off-centre on the lathe which is rotated against the knife. The effect is a variegated veneer.

Jaguar cars

The use of black walnut veneer on the dashboards and door panels of motor cars, most famously Jaguars, but also Bentleys, Rolls Royces and Austin Healeys, is most interesting. I was privileged enough to visit the Jaguar Veneer Manufacturing Centre near Coventry where I spent a few instructive and happy hours with Dave Adey and his supervisor Nick Pestley.

David Adey says that the first person to consider using walnut in Jaguars was Sir William Lyon, the founder of Jaguar cars. He was clearly a man of great taste and imagination, but no doubt, even he could not have predicted that the walnut veneer would still be put into cars in the 21st century. The name Jaguar was not the car's original name. The first models were called the SS (Swallow Sidecars) and were built in Blackpool. After the Second World War, the SS name wasn't acceptable because of Nazi connotations and that is when the famous Jaguar name was born. David Adey says that the walnut veneer has been used in all cars from the X type, the S type and the XJS.

The veneer they use is bought in Milan, Italy, but its original source is the United States.

They are looking for the finest quality burr walnut from the *Juglans nigra*, the black walnut. The burr is described as a deformity in the graining, which produces

Walnut burrs before being made into veneers.

fine patterning when created into veneers. The walnut burr is found at the base of the tree where the trunk meets the root. The deformity gives the image of a bulge or rubber tyre shape under the bark. The American walnut production involves sophisticated husbandry in as much as the nut production of the tree is deemed uneconomic after about 70 years. At this time the tree is felled for its timber value. The burr is cut away from the trunk with a chainsaw and the roots are also cut away. What is left is something that looks like an elephant's foot but considerably larger depending on the tree and how well it has grown. These elephant foot-looking chunks of timber are shipped to Italy. The burr is boiled for three days and by the time it has been through this process, the water in which it has been boiled is a black colour and I am told very smelly. After the boiling the bark, and as much of the sapwood as possible is taken off the burr using an adze (a tool which is the flattened side of a pickaxe). The burr is then put on to a lathe by spiking the trunk end securely. This enables the burr to be turned at speed. As the machine turns, a blade, rather like an apple or potato peeler, slices the veneer off. The depth is only about 0.48 of a millimetre. Each slice looks the same and this is the benefit of veneer. A symmetrical pattern or exact mirror image can be created by turning over alternate slices.

The veneer sheets arrive in the United Kingdom in what are called bundles on pallets. They are about one and a half metres long, a metre wide and about half a metre tall. They are wrapped in plastic to make the veneer retain as much moisture as possible. Once unpacked the veneers are inspected for the quality of the burr. That is to say the pattern of the burr, the amount of sapwood in the sample and the number of holes formed from weaknesses in the burr. Holes do not mean that the veneer is useless as they can be filled. The Veneer Manufacturing Centre say that

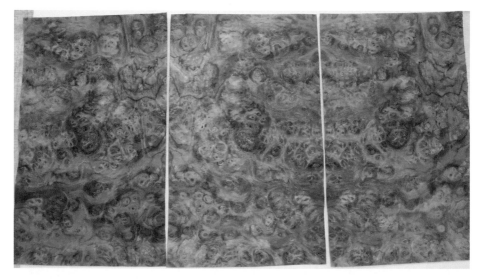

Examples of walnut veneers before being matched up for the interiors of Jaguar cars.

they do not buy the best quality veneers which are the most expensive and go to the furniture makers, but they certainly seek the second grade. The better quality sheets go into the more expensive Jaguar models.

The sheets are watered to maintain their moisture content and then go to an expert in matching the sheets to create book match. He will tape them together and then hand them to a machine operator who will sew the sheets together with nylon cord. The machine itself is to all intents and purposes a sewing machine and it gives the join some strength for the processes to follow. It is essential that the mirror image and symmetry is maintained and the sewing is said to be an improvement on the brown tape used in the past.

The sewn and matched veneer moves on to the cutting area. This is where modern technology really begins to appear. The cutting is done by using a laser cutter, which assures accuracy. The other real benefit is that the cutting is quick.

The veneer set moves on to the pressing machine. This is a large machine capable of applying pressure of 10.55 kilograms per square centimetre at a temperature of 155°C. It is these machines which create the panels. The panels for the early Jaguars were glued onto plywood panels. The problem with plywood is that it cannot be bent or moulded into a shape. The veneer experts at Jaguar say they cannot curve the plywood panel to create a pleasing modern design. In order to do this the Veneer Manufacturing Centre have been using prefabricated aluminium panels which are light and can be curved. In modern cars the lightness of the materials which go into their manufacture are essential for fuel economy, and the speed of the car. The cut walnut panel is backed by what is called tulipwood. For the manufacture of Jaguars, they use poplar. I was surprised by this revelation because I had always thought poplar was used for tea crates, coffins and matches. The dampness of the wood is what makes this wood ideal for these uses, but it is not the dampness that encourages VMC to use it. It is cheap and fits the process. The tulipwood is itself a veneer and therefore there is more than one layer. In order to make up the panel each layer must be laid in a different direction. That is to say that the grain of the tulipwood with a horizontal grain has on top of it a tulipwood with a vertical grain.

When the sets that make up the dashboard come off the press the edges are rough and have to be smoothed off. The important matching of the veneers is again checked. At this point the colour seems to be grey and the patterning of the burr does not show up that well. After the next process the colour is about to change.

The veneers in their sets are now heading for the veneer lacquering area. This is an area of high technology, namely robots. The veneer needs five coats of lacquer. The sets are placed on trays which pass under the accurate spray of the robot. Then, on a conveyor belt proceeding at a set speed the tray comes round for its next coat. When the veneers have dried they will be a different colour. The patterning is much more pronounced and black. The outer shell is not smooth and that is where the

Walnut veneer in a Jaguar. Notice how the two sides of the car match.

next stage begins. The trays move to another robot area. Here they are sanded and polished until they are perfect and ready to be packed for the main Jaguar or Land Rover factories.

There is two metres of dashboard in a Jaguar and that includes a map drawer. Taken from the centre gauge, the veneer is book matched to each side. Jaguar have their own quality control and the veneer goes through some very rigorous tests.

The finished product in the cars is a work of art in its own right. In an age of plastic and other synthetic materials, walnut dashboards and door panels might be considered out-dated. Yet, nobody could be unimpressed by the workmanship in the classic cars that dates back to before the Second World War. A prospective buyer of a Jaguar is confronted by a considerable number of different wood types for their dashboard. The best is olive ash, an attractive light coloured wood with a black distinctive grain. There is burr elm which no longer comes from the United Kingdom as our elms have sadly been destroyed by Dutch Elm Disease.

I have named other choices, but the best of all is the burr walnut and it seems that its popularity still prevails. I asked Dave Adey why this dashboard is still popular. He says that people who buy quality cars like the Jaguar look for high quality products inside them. It is like putting antiques in old houses because they look right whereas modern cheap furniture looks out of place. Walnut looks as right in a classic car as it does in a very modern car. The designers have adapted dashboards to keep pace with technology, safety and modern requirements, but the walnut or other burr gives it a distinguished and special look. Walnut is a product which people know as an important part of the Jaguar story, they know it fits well into the car and it is a distinctive wood with character. Cars can have a plastic look. My father often used to say of a car that it is "tinny". The leather and burr give quite the opposite look. It would be a fair prediction to say that the buyers will continue with the walnut veneer well into the future. The designers may want to change the interior, but they can only do this if the buyers want to make that change.

9

The walnut and the manufacture of guns

Hunting is a substitute for war. Guns and weapons were made over the centuries in order to satisfy the demands of war and hunting. The weapons for war have been developed for rapid firing, lightness and accuracy. Technology and mass production of military weapons has lead to a divergence from the sporting gun whose mechanism has not really changed much for a hundred years or more. Another difference is in the look of the sporting gun. The artwork in the engraving on the mechanism and the choice of a well-figured stock are essential to the sporting gun.

Walnut is famous for its use as gunstocks. Guns for hunting were first invented at the time of Henry VII. Before then the hunter might venture out with his "cadger" on horse back and indulge in the sport of hawking. Hawking is said to be the oldest field sport dating back to China, some 400 years ago. Bows and arrows, a later invention, could be used for rabbits and possibly deer, but would be a difficult way to hunt birds, even on the ground. Spears would be almost impossible to hunt birds with. Crossbows would have been used for birds on the ground and other game such as deer. There are two pictures in the Prado Gallery Madrid by the German artist Lucas Cranach the Elder. It is entitled "Hunt in Honour of Charles V at Castle Torgau" and shows a somewhat gruesome image of hunters with early crossbows hunting deer. It was painted in 1544. Cranach went to the Netherlands in 1509, painting Emperor Maximilian and his son Charles to whom the painting of the hunt was dedicated.

In order to kill a bird, the hunter would have used a small baked clay ball as the crossbow trajectory. This is the same kind of trajectory as the musket ball even used in the early guns. Henry VIII had very early guns. In 1540 he bought a German gun that was about a metre long and weighed three and a half kilos. This is heavy. It had a walnut stock and a Tudor rose carved on its side. The second gun appears in a 1547 inventory and is described as a "long chamber piece with fier locke sell in walnut tree".

The word stock, which is used to describe the back of the gun, is derived from a German word meaning tree. Early stocks were long poles placed in the ground and

Hunt in Honour of the Emperor Charles V near Hartenfels Castle, Torgau, 1544 by Lucas Cranach the Elder.

An early crossbow in the Brera Gallery, Milan.

attached to a cannon like device. Later guns were built with short stocks, some of which were designed to be placed on the chest for firing.

Walnut is a remarkable timber which is resilient to the pressure from the gun being fired. We learnt something else about its resilience from an article in the *Daily Telegraph* in January 2015. Someone had found a Wild West model 1873 Winchester rifle which had been left propped up against a juniper tree in the Nevada Snake Mountains for 100 years. It was rusty and the stock was in bad condition but still in place. It is not known who owned this weapon, but the ability of the walnut to withstand the changing seasons over this long period is further proof of its ideal choice.

I asked an expert the reason why walnut is used in gun making as opposed to any other timber. He was not sure why, but thought that its use had arisen by convention. In other words, it has strength and is easily carved so that the mechanism can be fitted.

It is obvious that in the construction of a shotgun, the two barrels have to be straight. They have to be in order to fire the cartridge. Their construction should be undertaken by a specialist blacksmith. There is a walnut fore stock for the front hand to hold. There is then the stock which is not necessarily straight because it is shaped to suit the shooter's eye. The stock is carved in such a way that one end fits into the shoulder, and at the front end the mechanism and barrels are at eye level to enable the shooter to connect with a bird on the wing. It is known that this concept took some time to develop and initially birds were shot on the ground. Henry VIII would have hunted in this way. It was understandable to shoot birds on the ground at that time because it was the only way to hunt successfully. Early stocks did not have the fishtail shape they have today. At the front end, they needed to fit the mechanism which fired the gun. The back of the stock was short and stumpy. The French makers of that age called their stocks a "Petronel". The translation is a reference to the breast from where the gun was fired. It would have been uncomfortable and painful.

By the early 18th century the flintlock had been invented which enabled the shooting of birds on the wing. In the 19th century a Scottish clergyman Alexander Forsyth invented the smokeless ignition which is how modern cartridges and bullets are made today. The father of development in shotgun technology was Joseph Manton who built the first double-barrel gun. He was an apprentice gunsmith in Grantham in 1780. He went to London in 1781, working under his elder brother and finishing his apprenticeship in 1789. He decided at that point it was time to work for himself. By 1831, with a shorter barrel, the art of shooting a bird in the air became popular. Manton took on apprentices who became famous gun makers in their own right. Lancaster, Purdey and others all went on to build their own guns and some gave their names to famous companies. Despite his skill and the number of guns he

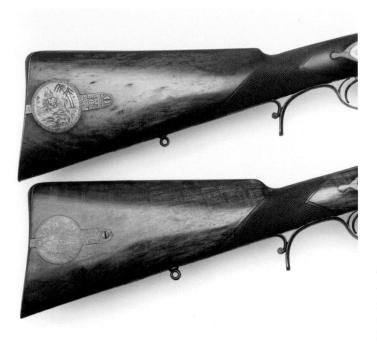

These double rifles were made in 1850 for Lieutenant Colonel John Jacobs.

produced a year, Manton's business did not flourish and he was made bankrupt in 1826. He died in 1835.

Manton's apprentices were about to enter a new era. Prince Albert and other notable people were to make driven game a popular sport.

Military weapons

The manufacture of guns as military weapons where walnut was used, put much pressure on the supply of walnut. John Evelyn talks about the shortage of walnut in France for the building of muskets and states that a sizeable tree could cost as much as £600. In 1806 the French needed 12,000 walnut trees a year to make muskets (*The Trees of the British Isles in History and Legend* by J.H. Wilks). In short, walnut was important in the "execution of war for many centuries". Fortunately, the wars in Europe saw more destruction of buildings than the forests or plantations in which the trees grew; otherwise we would not have the fine walnut plantations in the Dordogne and Grenoble today. In America, during the Civil War, many guns were built from black walnut. When soldiers were about to march, they "shouldered their walnut". (*James Purdey & Son* by Donald Dallas).

The Brown Bess was the musket used by the British Army between 1795 and 1815. Three million weapons were produced, originally built by the East India Company and later by Birmingham gun makers. With the French revolution and a war with Britain in the Spanish Peninsular for several years, all trade had stopped which

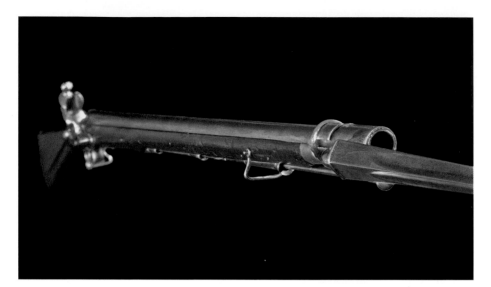

A Brown Bess musket with a bayonet attached as used at the Battle of Waterloo. It was stocked from Italian walnut to about ten centimetres from the end of the barrel, which was a metre long.

meant walnut stocks could not come from France and therefore another source had to be found. Supplies from Italy began to fill the gap although some experts have said that the walnut was not as good as the French equivalent. It was described as plain and "the wood was of an inferior heart and sap".

An 18th-century military carbine displayed in the Brera Gallery, Milan. Notice the rounded end of the stock made of steel so that it fits into the soldier's shoulder.

Inevitably in this modern and fast moving world, technology has moved on and walnut has some competition. Some rifles are now sold with stocks made of plastic, synthetic, graphite materials or carbon fibre. These are strong materials which enable stocks to be made quickly and cheaply but they do not have the same style. They certainly do not have the definitive colour or marbling that a walnut stock can provide, but they are practical. The disadvantage is that they do not absorb the impact of the shot in the same way that a walnut stock is known to do. Soldiers need light, practical and effective weapons and governments need the price to be low. In contrast, the fine modern shotgun makers of London are looking for a light, practical but robust gun capable of withstanding immense pressure. They are also looking for a highly artistic and impressive hunting gun for their wealthy customer. The less wealthy customers is looking for an ornate but cheaper fitting gun. Whoever the buyer is, the shotgun makers will use walnut because it is light, but strong and has a distinctive grain.

The walnut blank

Blanks are cut from the tree into logs, possibly up to a metre and a half long, and are up to a metre wide if they are from Turkey. The sawmill cuts them into cants or planks and then into blanks which is the rough shape of the stock. The blank is eight centimetres wide and around 60 centimetres long. At the cheek end, the blank is 18 centimetres wide, and at the fore-end, eight centimetres. They can be up to 90 centimetres long and at this length with a 38-centimetre width at the cheek, will make two stocks. It seems that there are difficulties in sourcing stocks for pairs and trios. It is not easy to find stocks that match, but it can be done and the pair normally comes from the same blank. The stock maker is looking for the finest grain flow. The best cuts come from where a limb is cut off the trunk or from the root ball. The root ball or log from the trunk is cut into cants and then put into what is known as sticks whereby the wood is stacked but separated with small pieces of wood to allow air to circulate around the cant. The timber is left in this state in a dry location for three years to allow the moisture to decrease. It tends to dry out adequately in approximately one year per two and a half centimetres thickness. One of the problems with air-drying is that over time the wood will split at the end. In order to stop this, the end is carefully sealed with paint.

The blank needs to have a final moisture reading of 6% to 8%. The drying process will continue when the blank is cut from the plank in store. Some stocks could be in store for up to a generation before being carved into shape.

The same drying effect can be achieved by kiln drying. This is where the blank is put into a heated oven until the desired moisture content has been achieved. This is not necessarily the right way to dry the cant and some experts maintain that the stocks are brittle. Mark McCarthy, a stocker at Purdey's, says that kiln drying is quick but tends to leave the inside of the wood a bit like a sponge when all the water has

A French walnut stock. It is well figured with a dark colour and is an exhibition piece.

been extracted. He prefers the air-dried method.

French walnut was commonly used for gunstocks, but many of the useful trees were cut with some urgency to fulfil the need for more guns during the First and Second World Wars. Purdey's say they were using French walnut until the 1980s.

Tony Kennedy, a gun dealer with many years experience says that 35 years ago he was able to buy good quality golden-coloured, strong French walnut stocks. He now says that the supply he obtained from the Dordogne has dried up.

The Turkish walnut

Turkish walnut is now said to be the most prized of all blanks. That country certainly has the reputation for having some of the oldest and largest trees. It is probably true to say that the older the tree, the better the graining. Stockers say they like the blanks to be from trees over 100 years old, but some blanks could be taken from trees as old as 300 years. Weather conditions over many years will put stress on the whole tree from the root upwards and this is said to improve the graining. Furthermore, the bigger the tree, the more stocks can be made.

The climate in the Turkish region near the city of Van in the south-east corner of the country, near the Iran, Iraq and Turkish borders, is ideal for the common walnut *Juglans regia*. The trees are found in an area of 700 kilometres diameter from Van. The summers are hot and dry. A local agent told me that the heat and sunlight gives the timber its colour. "The tree must see the sunrise" the Turkish agent said. The statement had a rather romantic sound to it as one imagines the mountainous area of eastern Turkey in the early morning, although I suspect it is more folklore than science.

The winters in that area of Turkey are cold and snowy. The agent told me his theory that the cold freezes the sap in the tree and this stress has the effect of patterning the timber. There is a little rain in the spring months. The climate is said to be a "steppe climate" with wide variations of temperature between night and day. The average temperature is 23 degrees centigrade in the summer and minus two degrees in the winter. The temperature can rise by 10 am in the morning to 35 degrees. That will have a severe effect on the tree as it can burn the leaves. To some people this may seem idyllic for summer living, but a little too harsh in the winter. For the trees it is quite harsh and we can assume that the walnut has adapted to survive in this varied climate.

The trees grow on the mountainside. I was told that the ground is often stony which is important. It is said that the colour of the blank will depend upon the minerals in the soil. Copper will bring out a red stock. (*The Shooting Times* article August 2015 Alec Marsh). Further north, the walnut is milky yellow because lodes of copper are not present. The Turkish walnut does not grow well near water. The assumption from this is that they get enough water from early spring rain and the thawed snow. Turkey has adopted a conservation policy with a replanting programme.

In Europe, where the trees are grown for their nuts, the distances between trees are seven to ten metres. I asked the Turkish agent what the ideal distance was for the best trees to make gunstocks. He said that he would expect a 300 year tree to be at least 20 metres from its neighbour. At 300 years old the tree has lost many of its larger branches and the reduction in leaf is also said to influence the colour inside.

The agents select trees in the autumn and might fell up to 400 per year. According to Alec Marsh's article in *The Shooting Times* in August 2015, some 35,000 have been felled in 35 years. The trees are individually selected from the plantations. The eye and expertise of the buyer, trying to imagine the colour and patterning inside the tree, is the way selection takes place. Often the selection is from gnarled and ugly-looking examples. The 400 trees will produce 2,500 tons of walnut which is about 75,000 blanks for shotguns and rifles. The blanks are graded from standard lower or medium grade to exhibition standard. The lowest standard grades are two or three. The medium grades are four to five and the exhibition grades are over five. The agents are looking for trees with no knots as these can produce weaknesses in the stocks. There is plenty at stake at the time of selection and the choice has to be right. The agents say that they are very careful in their selection and reject quantities of trees.

The felling of any tree in this eastern region of Turkey needs a permit. As a conservation measure they plant 4,000 walnut saplings per year to ensure the long-term regeneration. It is thought that this measure will not be enough to conserve stocks in the short term and that there could be a shortage at some point.

An exhibition-grade blank from Van, Turkey. Water has been dropped on part of the stock to show off the grain. The way in which the smoke plume-like grain is formed is a mystery of nature, unique to each individual stock. A finely-grained stock is a work of natural art.

The trees, when cut, have a large girth if they are 300 years old. They need careful handling so that the important and most valuable parts of the tree are not damaged. The weight of them must not be underestimated either. A digger is used to dig around the roots to a depth of about one and a half metres and the roots cut off underground. The branches are also cut off and useful wood sold to the local furniture makers. The length of the trunk from the bottom of the burr to first branches could be three metres. The cutting operation may take up to two days.

Back at the sawmill, the Turkish method of processing the blank is described by the agents. The trunk is put on a bench and cut into planks about eight centimetres thick. A steaming process follows which removes the water and the hardness of the wood is improved. The timber is then air dried for eight to twelve months.

Not all blanks will make the required grade. Whilst the agent has done his best to select trees which provide blanks with a whitish grey colour and fine straight black patterning, they still have to sell their product to the gun makers around the world. About 90% of the blanks sold are used for shotguns and 10% are used for rifle making.

Purdey is one of the best names in the building of shotguns. They tell me that they buy their blanks from the south-eastern corner of Turkey. They do not roam the area themselves, but rely on agents to source their supplies. Purdey say that they might pick 20 to 24 blanks from 250 samples. It could be that the grain does not look right. Their expert eye might consider the stock to be weak. They might say that the pattern in the blank is not distinctive enough. These are important points when making the choice for a blank, but there may be less obvious issues to consider. There may be cracks which may not be noticed by eye. When the wood is worked upon, it may be found to have what is called "shake". This is where the rings, grain or parts

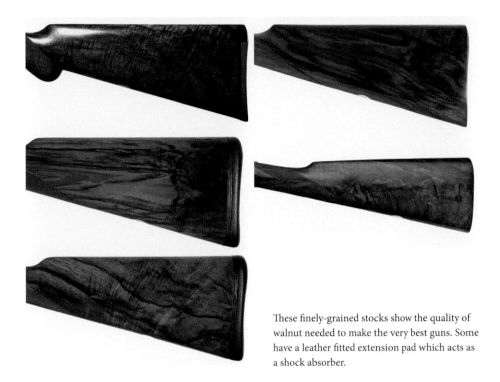

These finely-grained stocks show the quality of walnut needed to make the very best guns. Some have a leather fitted extension pad which acts as a shock absorber.

of the wood are weak. When the stock is worked, the wood literally falls apart at the weak point.

The gun makers say they tend to deal only with a small number of walnut agents with whom they have a relationship, so the dealer knows what the gun maker wants. Like any business relationship it relies on trust.

The colour of the stock is important, but that is a matter of personal taste. Some gun buyers like a dark stock and others prefer a lighter-coloured stock. Personally, I prefer the darker stock. The grain is the real selling point and is the mark of the gun's quality. The discerning gun buyer will look at the marbling. The black squirrely lines and shapes that form the grain of the wood make the gun into a work of art as well as a weapon. The job of the stocker is to preserve marbling, highlight it and carve the stock in such a way that it has a distinction.

The Spanish gun makers Aguirre y Aranzabal, more commonly known as AYA, have been making fine guns since 1915 in the town of Eibar in the Basque country. The guns range from the affordable to the "deluxe". The affordable guns such as the number 2 or number 4/53 are fitted with grade 2 stocks with 14 coats of oil. These are attractively-coloured with linear figuring, the brochure says. The "deluxe", number 1, number 37 and Augusta are fitted with an exhibition stock. This has fine colouring and the very best figuring. These finest-quality stocks are the most expensive and they have 35 coats of oil. What comes out of this extensive range of products is that

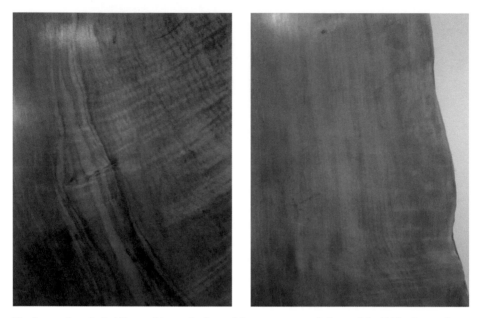

The fore stock or the holding position at the front of the gun must match the stock itself. The fore stock is cut from the front of the blank.

if the mechanism and the gun is of the best quality, the stock must be the same. The affordable guns are not fitted with the top quality stocks, but they fit the pocket of the buyer they are aimed at. This company obtains its stocks from the south-eastern corner of Turkey and from Armenia.

The French stocks tend to be bland and the colour is much lighter. Mark McCarthy says that the French walnut is easier to work and does not blunt the tools so easily.

Tailor-made stocks

Purdey say that one of the details they will consider is whether a stock has a straight grain in the part of the stock where the shooter holds the gun. The reason for this is that from their experience it provides some evidence that the stock is strong. The pressure from the shot will pass through the stock where the shooter's hand is placed as it passes to the end of the fishtail shape and the shooter's shoulder. This is the weakest point as it is where the stock is the thinnest. Purely for safety, this is where the stocker must be confident that the strength is there. I have seen a gunstock where a small chip of walnut shot off the stock past the eye of the shooter. He was unharmed but it was a near miss.

Paolo Zoli, an Italian gunmaker, says that when they make a gunstock, they will take into account the weight of the barrel, bore, diameter and more personal details like the shooter's height, weight, build and length of arms. The object is to try to ensure that the gun is mounted in exactly the same way every time it is fired.

The journalist and writer Mike Yardley comments that he likes the gunstock to fit snugly into the shoulder. He suggests a well-shaped curve at the shoulder end of the stock. The length of the stock is crucial. I was always told that the rough guide to see if a gun fits, is for the stock to be rested from the crook of the arm along the forearm to see where the trigger finger touches the trigger. This is an unscientific approach. The comb or the top of the stock should fit comfortably into the chin. There are ways of extending the stock if the gun needs to be refitted but it very difficult to match the grain of the extension with the original stock. That said, I have seen extensions where the joint is hard to see.

A stocker making a pair of guns for a grouse or pheasant shooter has to make the stocks identical. Some stockers will put in palm swells on the pistol grip so that when guns are handed to the shooter by a loader, his hand grips the stock in the same place.

The front of the stock is where the shooter's hand grips the gun. The stocker can make this to suit the shooter, but he needs to take into account the size of the shooter's hands in order to gauge the size of the gripping area.

The stock can have various shapes. The styles include a pistol grip and a deviation of this style, the Prince of Wales grip, invented for the prince in the 1920s. The pistol grip is ideal for the single-trigger shotgun mechanism, which has become so popular. For the double-trigger gun the straight stock is preferred according to Tony Kennedy.

The stocker will start by carving the stock to fit the metal working parts of the gun; the triggers, the safety catch, the side locks and the top lever. This carving takes up about one third of the time it takes to make the stock. Mark McCarthy at Purdey's says that as he carves, the wood adjusts as if it is alive and wants to protest at its treatment. Yet even if the wood wants to move, the metal parts have to fit exactly.

Walnut timber is capable of being bent or carved to make it look bent. The bending is essential to ensure that the gun is mounted correctly so that the eye goes straight down the barrel rib. If the buyer of the shotgun is spending a considerable amount of money he will expect the gun to fit exactly. Which way the cast goes depends on the shooter's eyesight. Mike Yardley says that someone who is right-handed with a dominant right eye will need only enough cast to keep the shooter looking down the rib. Where one eye is clearly dominant, the cast probably needs to be more pronounced. In my case I have a left dominant eye and shoot off the right shoulder. I have a small cast to the right but have to shut my left eye. Other people need a more pronounced shape stock.

Any stock is shaped in a downward angle from the barrel. This enables the barrel to be brought up from the shoulder to the level of the eye. When the empty gun is mounted the fitter should be able to see the eye just above the breach.

Stocks can be cast up, down or sideways. If the stock is to be bent to alter its shape, heating and oil are used. This is a difficult process and even the robust walnut

This curiously shaped stock is known as a crossover. The photograph shows a Purdey double barrel shotgun built in about 1945. These stocks are made from one (about twice as wide as a usual) stock blank and carved to the shape required by the buying shooter. In many cases the buyer will have been injured, possibly during the war. It could have been an eye injury or an arm injury which required this change from a normal shaped stock.

can crack or break. According to Mark McCarthy the bending process is not reliable. If the cast is required to a quarter, then it needs to be bent by a half. In other words the stock will try to bend its way back to its original shape.

Purdey do not bend stocks for new guns, they carve stocks to fit the shooter. The method is to find the central line at the butt of the stock. A second line to the left or right of the central line is drawn. The experience and skill of the stocker shapes and carves the stock to create the cast required. It can take a day to carve the stock and it is interesting to note that Mark says it takes longer to shape a stock from Turkish walnut than it does to shape French walnut. He also says that the Turkish walnut will blunt his tools quicker.

Guns should be a work of art. The top gun makers say that a gun finely made with an exhibition grade stock is a principal requirement of the person trying to buy a quality gun. Making guns with an ornate style has been practised since the 16th century. In the 19th century, shooting was an exclusive sport for the rich. Owners might own a suite of guns. A shotgun, a pair of double barrel pistols and a light pair of pistols perhaps for self-defence. Guns were also given as gifts. Some of these are ornate and made by master craftsmen. They are items to show off to your friends rather than try to shoot at a bird with.

On the fore stock and the hand grip of the stock the stocker will engrave chequering. This is important because it enables the gun to be held firmly, particularly

This sporting gun is displayed in the Royal Armouries in Leeds and it is called the Piraube. It was given to Charles Lennox, Duke of Richmond, by Louis XIV. Possibly during his visit to France in 1681/2 or 1685. (Charles Lennox was the illegitimate son of Charles II.) It is engraved with silver medallions with scenes of Louis XIV in a chariot with a sceptre topped with a Fleur-de-Lys and is inlaid with scrollwork in silver. It is thought that the designer was Jean Berain. (*The Field* May 2015)

A sporting gun built by Joseph Manton

A highly ornate Snaphaunce pistol made in England in *c.*1600.

when the shooter is in action in the rain. Carved in a pyramid shape ending in a point. The carved lines are close and Mark McCarthy says there are 25 to 26 lines per 2.5 centimetres for a .410 and 28 to 29 for a 12 bore.

The balance of the gun is important. A gun that is heavy at the front will be harder to mount and indeed to control. The stocker can adjust the balance of a gun by removing wood from the inside of the stock. The theory is that the weight of the barrel should be matched by the weight of the stock. I have a gun with two barrels. One barrel is 66 centimetres and the other 74 centimetres. The stock is 40 centimetres long as I am a tall man. My shooting, never perfect, deteriorates when the shorter barrels are fitted on the stock. The balance is different; I believe the barrel is lighter than the stock. It might be asked why there are two barrels if the balance is so different. The long barrel is for the high pheasant, pigeon and duck. The short barrel is supposed to be more manoeuvrable for the jinking snipe, the fast and most sporting of all gamebirds the grouse, the sporting partridge and the twisting, turning woodcock. It is possible to adjust the balance by putting lead in the stock with putty according to Mike Yardley in his book *Gunfitting*.

When the stock has been carved and produced to the correct size, it is finished. The grain is raised using oxalic acid. This has a bleaching effect which visually enhances the contrast of the stock. The grain is raised to remove the softer mid-portion of the tubular grain structure, leaving the harder lignin layer of the wood behind. The oxalic solution is absorbed by the softer core and expands above the surface of the harder material. On drying, the softer core is left exposed and is removed by sand paper or abrasives until the expansion of this softer material has been removed. The stock is then coated with three layers of a mixture of boiled linseed and the root of the alkanet plant. This gives the wood a red base colour.

These stocks have been cut in half to show how the gun is balanced with lead weights.

The final stage is the application of an oil finish. The ingredients are boiled linseed, beeswax and terebine dryers. The applied coat is rubbed off using linseed so that only a small amount is left in the grain. Rottenstone, which is powdered weathered limestone, is then applied which has the effect of filling the grain. Slakum oil goes hard once it has dried. The oil finish of any wood is made effective because linseed dries to its own volume.

One of the most impressive shots I have ever met had lost his left arm, but managed to shoot only off his right shoulder. His gun was a twenty bore with a light barrel, but for the shot to succeed, the stock had to be adjusted so that the gun was balanced and easy to mount with one arm.

Walnut as I have said is adaptable, hardwearing and strong, but it is not indestructible. A gun dropped breaking the stock or a stock run over by a vehicle or a stock just in need of replacement, can be an expensive exercise.

It would appear a difficult problem to replace it, but it is not quite as difficult as it seems. The blank is purchased and can be copied to exactly the same size and shape as the original. The interesting thing is that, after the drama and disappointment of the damage, the gun can be completely redesigned to the owner's specification and indeed be improved. The quality of the stock might be better than the original maker's choice. It is a matter of what is affordable and what fits the gun. There are machines that can match the shape exactly. The traditional stock maker would say he can copy a stock and that is undoubtedly true but it is fairly obvious that the machine would make a cheaper copy. The old stock is provisionally glued back together. As one part of the machine follows the shape of the original stock, the other part carves the new stock in the exact image of the old. It is then hand finished.

A broken stock.

10   The leaf and the nut

The leaf

Yes. The leaf can be used. There is a recipe for making Walnut Leaf Wine. The ingredients are:

- 500g walnut leaves
- 1kg demerara sugar
- 500g of honey
- 1 teaspoon acid blend
- 1 teaspoon yeast nutrient
- 3.5 litres water
- Montrachet wine yeast

First place the leaves in boiling water. Stir in the sugar and the honey. The leaves are then strained from the water. The acid blend is mixed in together with the yeast nutrient. The next stage is to add the activated yeast and allow the mixture to ferment. The fermentation takes about five to seven days before it calms down. The wine is placed in an airtight container in a warm place where the fermentation will stop. The wine is then transferred into another sterilized container, air locked and kept in a cool place for six months. The final stage is to bottle the wine and keep it airtight for another six months.

This recipe is adapted from *The Women's Institute Homemade Wines, Syrups and Cordials* (1954).

- 50 walnut leaves
- 40g orange peel
- 1 litre Armagnac
- 1kg sugar
- 5 litres red wine

Macerate the mixture for eight days, stirring twice a day. Then sieve, bottle, and store in a cool place for three months.

The black walnut leaf can be made into tea. It also has medicinal qualities for skin disorders such as dandruff. The leaves are boiled and the juice applied to bandages.

Charles Spencer in his book *To Catch a King* writes about the escape of Charles II from the roundheads during the Civil War. There is a description of the King staining his skin with walnut leaves to make himself appear tanned like someone who worked outside, not a grand prince. Knowing how staining from the juice is hard to remove, he could have kept the fake tan for a very long time, if not for life.

Charles was not the only person to make use of walnut dye as a disguise. William Watts MacNair, a cricket-loving surveyor, disguised himself as a Muslim doctor by smearing himself with walnut juice to escape into Kafiristan in 1883. Kafiristan is now known as Nuristan and is a district in Afghanistan. It was very dangerous at the time so saw few foreigners. This daring trip and William's disguise is thought to have provided an idea for Kipling's book *The Man who would be King*. The book was turned into a film with John Huston, Sean Connery and Michael Caine.

The nut

After the flowers which, as has been explained earlier are barely visible, comes the fruit or in this case what is commonly referred to as the nut. In fact, a walnut is not a nut, but a drupe. The word "drupe" is Greek and it means an indehiscent fruit which has a fleshy outer coat which encloses a seed. Almonds, olives, grapes, dates, coconuts and coffee are all drupes. There is a point of difference here and that is that some of these drupes have fleshy exteriors which are edible by humans and other animals, and some do not. Dates, olives and grapes have edible exteriors whereas walnuts, almonds and coconuts do not. The digestive system can consume the fleshy part of the drupe and the seed passes through the animal's digestive system to grow away from the host tree. The coconut can be moved, but research suggests that water is the carrier for this drupe. The walnut is the reverse. The outside fleshy part is not edible, but this fleshy part is the natural dye. The seed is edible.

The media continuously mention the fact that humans can suffer from nut allergy. The peanut, which can cause many allergies, is in fact a member of the leguminosae family and not a nut at all. Tree nuts including cashews, almonds, Brazils, hazelnuts, macadamia, pistachios and even coconuts, can also cause allergic symptoms. Allergies are not specific to any of these nuts, but to all of them. The severity of the allergy to an individual type of nut will vary from person to person. The symptoms may not present themselves at the first exposure or consumption of any of these nuts, but may well occur for the first time with children. The nut acts as what is called an allergen and the symptoms present themselves because the body reacts to defend itself. It is not for me to make any medical comment on this allergy, but a doctor's advice should be sought if an allergy is suspected.

The walnut is the national nut of France. They say that the region around the Dordogne is the nut-producing area. The Grenoble area is for timber production. In the United Kingdom there are very few trees that produce edible nuts. Beechnuts are produced every three years. The tree is also valuable for its timber. There is also the hazelnut which does not have valuable timber. The sweet chestnut produces a nut with a prickly exterior and timber that is used for fence posts and hop poles among other uses. Walter Fox Allen compiled a pamphlet in 1912. He says that "one pound of walnut meat being equal in nutriment to eight pounds of steak." His research was done in 1912 and at the time he suggested that the walnut industry was competing with the orange industry in California.

Pickled walnuts

By about June 24th, the Feast of St John, after flowering, there is a small green nut. It is not mature, it is not very big, but it is soft enough to stick a pin through it. The ability to stick a pin through the immature nut is important. You have to make sure the nut is picked well before the inside hard nut is formed. I believe the weather has a lot to do with the formation of the hard interior. If the weather is hot the hard interior will form faster. We learnt this the hard way during the hot summer of 2017.

The best practice is that the green nuts at the bottom of the tree are picked for pickling. A mature tree will produce several kilos of nuts and thus a large number of jars, probably more than an average family could consume in a few years. That said, it is not difficult to make a supply of walnut pickle yourself. The process has been around for centuries and is said to have been invented in Persia.

The first step is to make brine. In order to do this, mix four tablespoons of salt to one gallon of water. Prick the walnuts with a fork wearing a pair of rubber gloves so as to keep the dye off your hands. Put the walnuts in the brine and sink them by placing a plate over them. One recipe suggests that after three or four days, drain the brine and dry the walnuts. In ancient times it was suggested the nuts should be soaked according to the length of the lunar cycle. That is 29.53 days and then it is suggested the pickles should be dried by sunshine. This makes the remaining water and oil evaporate.

The next step is to pack the nuts into jars. After that, the maker has options. The principle is to make a hot sweetened vinegar to pour over them. Do this by adding to the vinegar, 350 grams of dark brown sugar, 30 grams of root ginger, 1 tablespoon of allspice, 1 tablespoon of black pepper corns and 1 tablespoon of mustard seeds. The alternatives to these ingredients can mean including honey instead of the sugar and a series of other herbs. The next process is to boil the mixture for ten minutes and then strain it into the walnut pickle jars. Another additive is ruby port which will enhance the taste. Leave the jars to rest for up to six months when they are ready for use. This method of pickling will enable the conserve to be stored for up to four years.

Green walnuts in vinegar after two days. The black marks are where they have been pricked with a fork.

What to eat with pickled walnuts

Pickled conserves are sometimes eaten with cold meats or cheese. It might be eaten instead of chutney in a ploughman's lunch, with English cheddar or with cold pork or beef. The pickle can enhance any pies such as steak and kidney or game, or be liquidized with stock to create a simple sauce for roast beef or steaks.

I have heard it said that there is a greater popularity for pickled walnut in the north of the country than the south. In the north it is eaten with hake, a member of the cod family. It is the most important fish in South Africa where the largest quantities are caught. Hake is a migratory fish and there are shoals found off Ireland in the Atlantic. Hake has a mild taste and the flesh is soft. It can be eaten as a steak on or off the bone and can be deep fried in batter. This is a popular choice in the fish and chip shops of Ireland as well as west Yorkshire and Sheffield. The local population seems to think that the brine vinegar and walnut flavours go well with the deep fried mild-tasting hake. The pickle adds something else to the single flavour to be had from just pouring pure vinegar over this very English meal.

To add taste to puddings, pickled walnut can be added to praline or poured over ice cream. A sweet pickled walnut would go with meringues, flans or in chocolate cakes.

Alternative ways to pickle

If there are more windfalls after the pickling harvest or there are trees which are prone to bad squirrel attack, there is another way to preserve the immature nuts. There is a Balkan recipe, according to Kris Collins of *Amateur Gardener* which helps to solve the problem.

1kg green walnuts
900g sugar
900ml water
4 teaspoons lemon juice
1 vanilla pod.

First peel the green walnuts then soak in several changes of water over a couple of days and drain.

Melt the sugar in water and heat until there is moderate syrup. Then cool until tepid.

Add the walnuts and lemon juice and the vanilla pod. The mixture is brought to the boil and simmered for 30 minutes. Remove from heat and leave the walnuts in the syrup overnight.

Next day reheat back to the boil and again simmer for 30 minutes. The walnuts are then placed in sealed jars until ready for eating.

Walnut ketchup

900g green walnuts
6 tablespoons pickling salt
850ml white wine vinegar
1 teaspoon cloves
1 teaspoon ground nutmeg

The method for making walnut ketchup is to prick each nut six times with a large needle.

Walnut Ketchup in preparation.

Put the nuts in a two-litre jar. You then need to dissolve two tablespoons of salt in one litre of water and pour the water over the nuts. Cap the jar, store at room temperature and leave for three days. At the end of three days drain the water. You repeat the two tablespoons of salt and litre of boiling water twice more. After nine days drain the nuts and leave in a colander in the sun for two or three days so that they blacken. Place the nuts in a blender with vinegar and grind. The same process can be done using a pestle and mortar. Then return the walnut vinegar mixture to the two-litre jar and seal with a non-reactive cap. Leave the jar at room temperature for one week.

Strain the walnut vinegar mix through a jelly bag. Squeeze the bag to extract the liquid. In a blender or mortar grind the garlic with a little of the walnut liquid. Combine the purée with the remaining liquid in a saucepan and simmer for 15 minutes. Pour the hot liquid into sterile bottles and cap or cork them. You can store the jars in a cool dark place for up to a year.

In his *The London Art of Cookery* (1800), John Farley describes the making of Walnut Ketchup:

> Having put a quantity of walnuts into jars cover them with cold vinegar and lie them close for twelve months. Take out the walnuts from the vinegar and to every gallon of liquor, put two heads of garlic, half a pound of anchovies, a quart of red wine and of mace, cloves, long black and Jamaica pepper and ginger, an ounce each. Boil them all together till the liquor is reduced by half the quantity and the next day bottle for use.

I assume that John Farley means ripe walnuts, but he doesn't say so. He goes on:

> or take green walnuts before the shell is formed and grind them in a crab mill or pound them in a marble mortar, squeeze out the juice through a course cloth and put to every gallon of juice a pound of anchovies, the same quantity of bag salt, four ounces of Jamaica pepper, two of long and two of black pepper and mace, cloves and ginger each a quarter of an ounce and a stick of horse radish. Boil all together till reduced to half quantity and put into a pot. When cold, bottle for three months.

Walnut chutney

Osman Yousefzada describes a fail-safe recipe for walnut chutney.

225g coriander leaves
12 green chillies with seed removed
20g garlic cloves
25g broken walnuts

25g raisins
25g castor oil
2 teaspoons salt
4 lemons or 6 tablespoons vinegar

A food processor is required to mix the coriander leaves, chillies, garlic, walnuts and raisins. The mixing should be so well done as to create a paste. The next stage is to dissolve the sugar and lemon juice, which completes the task. The chutney is a good complement to rice and kebabs

Walnut wine

The green walnut can be used in a recipe for wine which increases my belief that the walnut is the most versatile of fruit trees. The Italians make a cordial called "Nocino" which is made from green walnuts. Modena is one area where it is made and another is in the area of Campania. Sorrento is in this area and in the surrounding area are gardens with olives lemons and some walnuts. The walnuts grown here are large in comparison to those grown in the United Kingdom.

The Villa Massa is situated on the edge of Sorrento. The wine making business was started in 1991. I was told that 90% of their production is the liqueur "limoncello". The Nocino is another line of production. The green walnuts are picked on the feast of St John, which is the 24th June, from trees owned by a single farmer and delivered in Jute sacks. The walnuts are cut into small sections and put into alcohol which is from a regulated agricultural source principally molasses. The tank is sealed for six months. When the tank is opened in January, water and sugar are added.

Walnuts ripening near Sorrento.

Some producers put in herbs and use caramel. In theory the liqueur is 50% walnut. Villa Massa say that they make the black liqueur to order. Very little of it is sent abroad to countries like the United States. Nocino can be found in Italian airports alongside the bottles of Limoncello. They say it is best served chilled after a meal.

There are many other recipes created by individual families and others. One method is that green walnuts are cut into quarters are then soaked in grain alcohol mixed with sugar, cinnamon, cloves, water and lemon rind. If it is made to this original recipe, it is 70% alcohol or 140 proof. It can be made to a lesser proof of 30% alcohol or 60 proof which is the Villa Massa recipe.

In Sarlat, it is possible to find other bottled versions of the nocino-type liqueur. One shopkeeper told us that the way to drink it is to put about two centimetres of walnut liqueur in a glass, add two centimetres of crème de cassis and top up with wine.

An enterprising Scottish liqueur maker, Angus Ferguson of Demijohn, took some of our green walnuts and experimented with various alcohols to find an ideal taste. Angus found the best combination was green walnuts with apple eau de vie. Julian Temperley makes apple brandy in oak vats and this is the eau de vie used to make the British walnut liqueur.

Walnuts and some of the Demijohn walnut liqueur – a British version of the Italian drink nocino. The nuts were taken from trees we grew in Sussex.

I tasted a rather good walnut wine made near Sarlat. The recipe was explained to me by a very helpful market lady. She told me that it is made as follows. They take 40 green walnuts and put them into one litre of Bergerac wine. She said that they add 20 pieces of sugar. By that it is assumed she means "lumps". One glass of eau de vie is the final ingredient to fortify the wine. The wine is stored for a year and then it finds its way into the market.

There is a French recipe for a wine made from the green walnut. Take 10 green walnuts, one orange, three and a half litres of 12/13% red wine, and half a litre of 40% alcohol. Chop the walnuts and orange into the wine, alcohol and sugar. Mix and stir. Filter into bottles and leave for 40 days. Lifting and turning the bottles from time to time will aid development in a cool dark place like a cellar or shed.

There is a German version called "Faude Feine Brände". It is made in southern Germany and the label says it is made from hand-selected green walnuts, pure spirit, herbs, spices and sugar. It is matured in Sauternes barrels for several years.

11

The summer and husbandry

In mid-summer months shoots appear beyond the growing walnuts and the trees turn a reddish colour.

The summer brings on a growth in the tree and new leaves appear. The pink colour in the green is a startling combination.

This walnut throws up new growth in July/August with a fine colouring from the pink leaves before they turn to vivid green, the usual colour of the walnut. The colouring lasts for about two weeks. Not all *Juglans regia* put on this show. In this photograph there are no nuts. We harvested them for liqueurs.

Ripening walnuts in the Dordogne. They are clean and disease-free.

Pruning for a better harvest

After the green walnuts have been gathered or even if they have not been harvested, pruning must be considered. Pruning is important to the tree and indeed to its management. Quite often there is no need to prune at all. One of the reasons for this action is to get the shape right in the garden environment. The shape should be like a tulip or goblet shape. Ideally I think the lowest branches should be about two metres from the ground. The French say that there should be eight main branches to give the tree the goblet shape.

My pruning has the aim of keeping the shape and preventing the trees from growing into each other. I will take out branches which are growing under a branch and which will grow downwards in time. I try to look at the tree and imagine its growth. I want to cut the twigs which are no more than two and a half centimetres thick. I believe that the tree will not bleed so badly with this kind of policy. There is also the squirrel and I want to make the tree difficult for them to climb and ultimately raid. I do, however, stop pruning after mid July.

Another reason might be to make it easy for machines to get under the trees for grass cutting and picking at harvest time. In order to pick the walnuts easily, the tree should not be allowed to grow too tall. If the trees are not shaken to make the nuts drop, long poles are used to 'bash' the trees to get the nuts to fall. If the tree is too tall the pole needs to be long and may be unwieldy to use. This is another reason to prune. We found that having branches that hang down almost to the ground is a disadvantage because as the weight of nuts increases as they mature, so the branch touches the ground.

Another way of looking at the pruning question (according to the RHS) is to try to train the tree not to grow up. This works if there is plenty of room. The way to try to do this is to cut the vertical leader and the side shoots should be plucked at the fifth or sixth leaf. I have not seen this done with walnuts in Britain or France.

There is a well-known rhyme which goes "The wife, the dog and the walnut tree, the harder you beat the better they be". This is clearly unacceptable behaviour in the case of the first two beatings but what about the third? Does it really make a difference? Putting any tree under stress encourages it to produce seed, fruit or nuts. In any event what does the beating mean? Does it mean lacerating the bark or a heavy pruning of the top branches? It seems doubtful. The answer is more likely to do with the harvest. The most basic way to harvest the nuts is to use poles and in a sense, the harvester has to beat the tree to get the nuts. I doubt this would make the tree any better from the point of view of its shape, as clearly the enthusiastic beater will break branches. It must also be admitted that this activity will put the tree under mild pressure.

John Evelyn in *Sylva* says that, "In Italy they arm the top of a long pole with nails and iron for the purpose [of harvesting] and because beating improves the tree which I do not believe more than I do that discipline would reform a perverse shrew."

There is care needed in this trimming because there is a right time and a wrong time to prune. Walnut trees will bleed if the pruning is done in the spring or late winter when the tree should be dormant. The best time to do it is in the late summer. In fact the recommended time is based on legend. The suggested date is the feast of St Swithun, which is July 15th. The day on which it is said that if it rains, there will be rain for 40 days. In my lifetime that 40 day statistic has never been achieved, but there have certainly been very wet summers. This is not a new theory; it was suggested in Elizabethan times.

The pruning here is caused by traffic. Not an ideal way to prune.

It has been put to me that the French will use a mechanical bush-wacker hedge cutter every three years giving the side of the tree a real beating. The bush-wacker is a flail cutting machine which will mulch the twigs into chippings. There is no top to burn or remove. The hedgecutter prune seems to be drastic but there is considerable top growth, a tangle of crossing branches and a rather confused growth. I believe hard pruning isn't tolerated, so this method is too hard on the tree and is not recommended.

One of the aims in the cultivation of the trees is to make picking of fruit or walnuts as easy as possible. Side growth rather than vertical growth is preferable for picking. One of the problems with the French walnuts is that they tend to produce tangled branches. According to Chris Brickell and David Joyce in their RHS book on pruning and training, any crossing branches should be pruned out. The *Juglans regia* walnut tree has a rounded canopy. It tends to have heavy branches with twiggy off branches. The shape is important to give a pleasing profile.

A French producer I spoke to in the Dordogne said that he tried to create a goblet shape of the tree. He says he never prunes right back to the trunk. He likes to prune the end of the branches that hang down, saying that this encourages the goblet shape. Where he has cut, he says the tree will sprout. He says he would never use a bush-wacker or any other mechanical pruner to 'beat' the tree. He will take hand loppers and carefully prune it.

There has to be a leading branch for growth. It is possible that by disrupting the leader by pruning may damage the tree. There is a school of thought which suggests you can take two interior upward growing branches and palmate them. Palmating trees is often used on fruit trees where they are espaliered or pruned into a fan shape so that they grow flat against a wall. Bearing in mind the buds are at the tip of the branch, there is a practical sense in this method. In France I saw one walnut plantation which had co-dominant leaders, that is to say there are two lead stems.

Frost is something we see in the United Kingdom and Europe and when it occurs late on in the winter, it can damage the tips of the branches where the buds and new leaves are. These need to be pruned and quite severely so that all the frost damage is cut out. The dead material is of no use to the tree and cutting it out will speed recovery during the summer.

David Murdoch writes in the *New Zealand Walnut Growers Manual* about pruning the trees and his findings from studies he has conducted since 2004. He says there are three methods. Hard pruning involves directing the leader back to the round wood. This usually means cutting below the closely-spaced nodes at the top very early in the season. The benefit of this is that it promotes vigorous shoots, which encourages long straight leaders. The terminal bud seldom grows in lightly-trained trees because another bud underneath overtakes the previous year's shoot. The disadvantage is that breakages can occur in the wind. Research indicates that the height and diameter are no different if hard or light training are used. Medium pruning is a method where branches are removed which exceed half the diameter of the trunk. Light pruning is a method where branches exceeding two thirds of the width of the trunk are removed.

12  The ripe walnuts

There are areas of the country where walnuts are not found. One notable garden and tree expert in Cornwall said there are few examples in the county because the climate is too wet. It is true that the wet climate isn't ideal for this tree.

As the nuts ripen, it becomes obvious that there is a need to make sure the ground is clear under the tree. The practical reason for cutting under the trees is so that fallen nuts can be easily seen. I use a mower with a 75 centimetre cut widening a circle from the trunk so that the grass is cut to just beyond the canopy of the tree. This enables nuts on the ground to be easily picked up.

Some nuts will begin to appear on the ground rather earlier than expected. When they are picked up, they seem light and hollow. The outside of the nut is paler than usual. These nuts are probably rotten inside. There may be signs of small holes on the outside indicating that weevils have got inside to consume the meat of the nut. The early fall of these nuts is a false harvest and there could be a few more weeks to wait before the real harvest begins.

A cleared piece of ground under a walnut just before harvest.

Walnuts on a tree in Sussex slowly ripening in the autumn sun with cattle grazing in the distance.

There is no doubt, picking walnuts needs a bit of patience. The walnut is not ripe until the outer shell cracks open. There can be signs of squirrels or other predators. They seem to know when the nut is almost ripe. The thief will leave the green outer shell on the ground, but no sign of a nut.

The real harvest is quite a sudden event. One day there will be no sign of any cracks in the green outer shell. Then suddenly, within two days, the outer shell can split into four quarters or the skin will split in a haphazard way with a rather jagged look to it. The first signs are that black lines appear on the outer shell and then the crack appears. In other cases there is just a clean crack. You see the same effect with the prickly sweet chestnut. Even when the outer shell has split wide open, the nut

A ripe walnut splitting open. Notice the dye on the green flesh.

may still remain attached at its head, Wind, tree movement and drying will eventually cause the nut to fall. I have noticed that the outer shell cracking open also allows the plum-like shape to be picked. The gentle twist of the apple tests whether the fruit is ripe. The same applies to the walnut, but only when the outer shell has opened. This whole process is nature in action and the ripening does not seem to like any influence by man. If the nut is picked too soon the outer shell is hard to peel and is very wet indeed. The light green flesh seems to cling to the nut itself and it is hard to get the nut away from the outer shell. That is until the nut is ripe and it is ready to shed its outer case. The moment it is ripe it needs to be picked. That way, the flavours can be maintained. Varieties can behave in a different way. One tree we have is very prolific, but the outer flesh can go black before the nut drops.

There are risks if harvesting is not done on time. The kernels may darken. There is a risk of insects getting into the nut. One is the walnut husk fly, another is the navel orange fly. *The Commodity Storage Manual* written in 1995 says that the time to harvest is when 85% of the crop is ready to pick.

Picking walnuts when they are ripe is more a matter of looking down than looking up. They are on the ground. At this stage, shaking the tree may well bring down the nuts which are ready. A long pole with a hook large enough to catch a branch and shake it is a possible tool. The hook needs to be about the size of a shepherd's crook. That is to say large enough to catch a sheep by its neck. I think care is needed with this device. The picker must to be careful not to break the branch in a moment of enthusiasm. At this late time of the year the tree would bleed.

During our harvest we had some help from our two dogs. The labrador would dutifully pick one nut up and bring it to hand. The cocker spaniel found them tasty and ate them. This is not a practical or necessarily a hygienic solution, but we found it added to the fun of the harvest.

I have explained earlier that not all shells have healthy nuts inside. The light shells and those blackened are normally empty. The outer shells that are green and healthy when they fall or are open are a different matter.

Quite quickly and soon after the outer shell layer is breached, the inside flesh turns to an avocado colour and then in time goes black. The green colour does have a use. This seems to be a natural dye and indeed it is. Walnuts contain a chemical known as phenol. It was turned into ink and used in the writing of early manuscripts. It was also used to dye wool, but over time it was found that the wool deteriorated. The dye is also of interest to the tattoo artist.

For the walnut picker, it has the effect that hands and fingernails get stained and the colouring is difficult to remove. For any walnut grower who has a smart party to attend, I do not recommend picking walnuts, as you will surely be embarrassed by the state of your hands. Surgical or rubber gloves are quite good for this work but I found that bulky gloves are not. If you pick off the tree, you need to get your fingers inside the green husk. Most farmers in France don't take these kinds of precautions, and are content to get their hands dirty and will not worry about the stain. Let us assume you have not taken any of these precautions, how do you get rid of the stain? A solution is to allow it to wear off, but that way the stain takes some time to remove. Even if the hands are washed regularly, the removal can take a week or two. There are some less obvious solutions. One suggestion is that the juice from a crushed red cabbage spread over the stain will cause it to disperse. Another way is use ordinary petrol. I am not sure the medical profession would recommend this, but it seems a more reasonable way to deal with the problem. The final idea seems to have come from someone with a sense of humour. It is suggested that cow's urine will do the trick. I have no idea if this method works and I must say I have a reluctance to try it. I am sure my readers will feel the same.

The number of nuts on the ground will increase day by day in the harvesting time. For the small producer, speed in collecting them is important. The back-breaking method would be to pick by hand. The photograph below shows a hand-held French tool which picks up nuts in an efficient way. This tool can be bought in the UK. The technique is to roll the device along the ground and the nuts will go through the wires becoming trapped. By dividing the wires the nuts can be poured into a container.

French walnut gatherer.

The ripening process if allowed to take effect in the natural way has disadvantages, particularly if the autumn is wet. We found this was the case in the UK. It is a fact that the nut has to be dry to keep for long periods. A fully ripe nut which is dry should not rot. It was noticeable that when the Equinoctial gales and autumn rain arrived at the same time as the harvest, the nuts we picked off the trees were in better condition and a bit dryer than those we picked from the ground. I suspect this is because in the tree there is air circulating but on the ground there is more dampness.

The nuts at this stage are vulnerable to the threat of theft from the squirrel. Apart from the warmer and dryer weather in southern France, animal predators in the UK are a fundamental danger in growing walnut. In France they do not seem to have animals or birds that steal the nuts. We are very vigilant at harvest time. We put out traps under the trees, using bread with peanut butter spread over it in an attempt to make a tastier meal than stealing nuts. Our routine is to check the trees every two days particularly when they are young and the crop is small. The crop from our grove began after the trees had been on the farm for three years. At that time we got about a bucket full off 50 trees. When the crop grows the squirrels cannot keep up with a crop which could be 50 kilos per tree.

The picking is quite easy, but cleaning, washing and drying is not. We remove any green outer layers and wash and remove other vegetation by hand. We dry the nuts on racks outside in the autumn, or above the Aga. It is not at all efficient, but it seems an effective way to do it.

Because of the size of our grove, it is not economic to have harvesting machinery. In a crude economic philosophy, I believe you would not have a sit-on lawn mower for a lawn six metres by six metres. In just the same way, there is no economic sense in having harvesting machines for a few acres.

I mentioned before, the damp atmosphere in the UK is due to our Atlantic weather systems and climate. During wet summers we found about one in eight nuts was completely black inside. If there is a cold spring and we see late fertilization of the buds, the harvest will be late. In the United Kingdom the autumn weather changes and there are wet days with some heavy dews at night. So it is important that the nuts are picked up off the ground as quickly as possible.

13

French and other walnut growers

The principal area in France where walnuts are grown for the nut is the Dordogne or Périgord region. It is an area of real beauty, with rolling hills, woodlands of chestnut and oak, pastures with Limousine cattle, maize, other cereals and magnificent walnut groves. These plantations are both large and small. The quality of the land is different in the river valley to that on hillside. As the ground steepens, the fields and walnut groves turn to forest. Yet in the hills you come across groves by streams and alluvial soil, but also elsewhere where the land is not too steep to allow machines to work.

It is rural bliss in which you do not feel there has been much change in the way of life, despite mechanization and technology. The mechanization has no doubt reduced the work force requirement for the walnut industry, but it is equally obvious that it has made the job of harvesting very much easier, more efficient and consequently has reduced the costs. Not all French walnut farmers have enough trees to be able to afford harvesting machines. They need to adopt the original method of harvesting with labour picking up the nuts by hand or have co-operatives across several farms to own machines. This concept of cooperation is something we don't seem to adopt much in Britain. There are share farmers and machinery arrangements, but farmers generally seem to want to work individually.

It would appear that under French law, a person does not have to own the land on which his tree is planted. Jeanne Strang says in her book "the division of peasant's estate often involved one member of a family being allocated trees in a field allotted to a relation." This kind of arrangement seems difficult to manage and largely depends on good relations between the owners.

The larger walnut plantations in France are planted in such a way as to allow mechanized cultivation and harvesting. Many of the groves are cultivated so that there is bare earth beneath the trees. Other groves have strips of sprayed bare earth about two and a half metres from the base of the trees.

It takes about ten years for the French tree to produce a full crop. Farmers say that the tree's crop will decline after about 20 years. The timber, I was told, is

A typical walnut grove in the Dordogne This plantation is close to a road. It is a neat and tidy scene with the trees pruned to a sensible height so that machines can get underneath and the ground is clear.

A younger grove with older trees on the outside.

not much good for furniture or gunstocks in the Périgord. It was interesting that nobody I spoke to gave a response as to what happens when the tree's nut producing life is finished. It would appear that a proportion of the timber is wasted. There are of course some exceptions. Some timber is used for household furniture and building.

The French aim is for nut production in orderly plantations. The growth of the trees is restricted by the closeness of the planting. In some areas the trees are thinned and we saw hay being made and other activities. A satisfying sight was a herd of Limousin cattle peacefully grazing amongst the trees. I thought this diversity interesting as the farmer was clearly trying to get what nut production he could from the older trees and an income from his cattle. I notice from the photograph below, that the lower branches have been cut off. This helps to prevent damage from the cows.

In the Périgord the walnut is common and there are numbers of trees on small patches of land in twos and threes for local home consumption. The local growing for oil or whole nut is typical of the small French farmer's rural way of life.

In France there are three stages in the harvest of the mature nut. The fresh walnut is harvested as soon as the nut has separated from the outer shell. At this stage the nut is very moist. The nut itself is white and the skin can be easily pealed. The time frame for the harvest is from mid September to mid October. Storage is a problem due to the moisture in this area we were told. The nuts either need to be

Limousin cattle grazing amongst walnuts. One of the benefits of keeping cattle near or under walnut trees is that walnuts do not attract many insects, whereas other trees do. It is thought that the scent they give off is a repellent to the insect, but not the cows who seek relief from flies in the warm summer months.

A narrow strip of walnut trees amongst corn fields.

The photograph shows a more extensive grove on the banks of the Dordogne opposite La Roc Gageac. The trees are planted neatly in lines and are well tended. The ground beneath them is bare.

A forest near Hautefort.

eaten quickly, preserved in a dry environment or transported away. The dry nut is collected from early October. At that time, the nut has fully ripened, lost some of its moisture, and dropped to the ground. It is washed and dried and the effect is that there is a red coloured nutmeat. It is possible to keep a dry nut for as long as a year. The final product is the 'Shelled Périgord Walnut'. This is the ultimate product used in patisserie, cake making and other confectionaries. The nuts are graded into sizes and by colour. They are also guaranteed by the AOC/PDO label. Appellation d'Origine Controllé of walnuts and the PDO label stands for Protected Designation of Origin.

There are six notable varieties in the Dordogne, Lot and Corèze departments. These are Corne, Franquette, Mayette, La Grande Dame and Parisienne. This region is the centre of nut growing in France. There are 7,000 hectares of walnut groves in this area which have a perfect soil type of calcareous clay. Calcareous soils may contain quantities of chalk, marl, limestone and most important for fruit and nut production – phosphates. The tourist guide says that there are some 2,600 hours of sun per year.

The area's weather is influenced by the Azores High and the Gulf Stream, therefore there is a warm climate. There is rainfall but not in abundance which is why the clay soil is better than a chalk soil. Chalk soils will not retain water whereas the clay soil will do so, a local farmer told us. There are wooded valleys with hills not rising above 500 metres. It would seem that walnuts are not grown further east and in particular not in the Rhône Valley, because of the famous Mistral. The Mistral is a wind that comes off the Alps and blows down the valley at some strength which makes walnut production difficult.

The Grandjean variety is a medium-sized round walnut only used for cracking. It has a fleshy nutmeat with a pronounced taste. This nut is grown in the area of Beynac, Daglan, Domme, Carlux, Salignac and La Roque Gayeac.

The Corne L'Isère variety has a hard shell. It has a sweet tasting meat with a fine texture. It will grow in relatively poor soil. The shell is rough, but it has a good kernel. It will keep well and makes a fine table nut. It is found near the Dezère river at Terrasson, Hautefort, Saint-Robert and Ayen.

The Marbot variety is a large-sized round nut with finely-veined nutmeat and is preferred for the production of fresh walnuts. It is found near Collonges-la-Rouge, Meyssac, Beaulieu-sur-Dordogne and Argentat.

The Franquette variety has an elongated blond shell inside its case and the walnut has a delicate taste. About 65% of the oil production comes from this variety. The nut makes up 41% of the nut. It is able to withstand the rough treatment from machines. This nut is found near Souillac, Martell, Carrennac, Bretenoux and Saint-Céré. It is also grown in the United Kingdom and indeed in my plantation.

The Mayette variety is a large pointed nut, and the nut makes up 45% to 50% of the whole nut.

The Parisienne variety is rounded in shape. The nut makes up to 35% or 40% of the whole nut.

There is quite a lot of research going on into new varieties. Fernette and Fernor are two varieties that produce late flowers, thus providing protection against late frosts that can so damage the crop. A better variety called Lara will in time replace Franquette. The French want to try to breed trees that will produce flowers and therefore nuts not just at the tip of the branch. The benefit of this genetic selection will increase the crop size, but could also lead to damage to the tree. The strength of the branch is limited and with a heavy crop, they could break under the weight. There are certain varieties of plums, such as the Victoria plum, which crops heavily and the result is that branches do break under certain conditions particularly if it is windy.

Grenoble

The Grenoble area is the other main walnut growing region in France. This area is to the east of the Rhône and out of the way of the Mistral. Grenoble is where the Rivers Drac and Isère meet. It is at the foot at the Alps and famous for winter sports. The city is 2,000 years old and was the centre of glove making. Historically in this area, the walnut could be bartered to pay the farmer's rent. Walnut crops were measured in Setiers which is a bag about the size of a pint pot. The increase of walnut groves occurred in the 1870s, principally as a result of a disease which killed silk worms and the arrival of that grape destroyer, Phyloxera. Walnuts provided an alternative crop following the demise of these industries. The main varieties are Franquette, Mayette and Parisienne. The region was helped by the coming of the railways which facilitated the transportation of the crop to other parts of France.

The Italian walnut

The walnut tree grows over much of Italy, but the famous area is Campania, a region in southern Italy. This region is dominated by the currently quiet volcano Vesuvius. It has not always been so quiet and has famously erupted a number of times spewing ash over a wide area. The result is that there are good potassium levels and other trace elements in the soil. There is another nourishing element in the soil and that is the residue from the sea which retreated many centuries ago.

Sorrento sits on the southern side of the bay of Naples with Capri a few miles offshore. According to the writer Diodorus (c.60–30 BC), the ancient city of Sorrento was founded by Liparus. It was probably influenced by Greek civilization and this may be why the walnut arrived in this area. The warm climate and the quality of the soil have had a beneficial effect on the trees. In this volcanically mountainous region there is extensive terracing, often crammed with lemon trees, vines, olives and occasionally a walnut tree. The Sorrento walnut is a *Juglans regia*. I saw plenty of trees near the twisting, mountainous road between Amalfi and Sorrento but they were not as extensive as in France.

Walnut trees precariously planted on the edge of the road between Amalfi and Sorrento.

Harvesting on a large scale in France

French harvesting in its basic form is waiting for the nuts to fall and then picking them up from the ground. We were told a French method is to use a fishing net with a flat edge. An alternative local method is to shake or beat the branches with a long pole.

These days the inventive minds of engineers have created a modern method of dealing with the harvest. A tractor with a front loader fitted with a tree shaker can be deployed. Alternatively a machine, albeit expensive, can be purchased which vacuums up the nuts with an airflow.

The revolving paddles sweep the nuts into a bed under the machine. Leaves and earth are sieved and removed and the nuts conveyed into a hopper at the back.

This machine was yet to be used to harvest walnuts. It cost €85,000. In order to make it cost effective it needs to harvest 2,000 trees a year.

The conveyor belt mechanization of the harvesting process and, on the right, the washer.

Once the nuts have been transported to the farm they are put into a barrel-like sieve and washed. The output of this machine is about 300 to 400 kilograms per hour.

The nuts are then rinsed, sorted and any green stalks removed. They can be air-dried. In a warm September this could take three to four days.

Quality is everything when selling any food product and there are some methods used to improve the taste of walnuts. The nuts can be soaked in milk. It is a natural additive that can take out the sour taste if the nut has begun to get old. If the nuts are not bright enough in colour, they can be whitened by soaking them in salt water.

Air-drying racks in the Dordogne. The cheap yet effective way to dry walnuts, but dry days are essential.

Walnuts can be dried in a specially built dryer. The dryer above is fuelled by wood. It has a series of trays that contain the nuts and wood is supplied at the bottom of the device. It can dry many kilograms of nuts at a time.

Storage

According to Walter Fox Allen who wrote a book on English walnuts, "In England the nuts are preserved for the table, where they are served with wine. They are buried deep in dry soil or sand so as not to be reached by frost, the sun's rays or rain or by placing them in dry cellars and covering them in straw." That was written in 1912. Storage methods have changed since then. The aim must be to preserve nuts for 12 months. The problem with walnuts is that they are rich in oil and if they are not stored properly, the oil will go rancid. The opinion seems to be that keeping the shells on, gives the walnut a longer shelf life. Dry shells will survive at room temperature for several weeks. The best way to keep them is to store them in an airtight container. The opinion of Walter Fox Allen, apart from the "deep burial" is still relevant except that it is recommended that they are stored off the ground and at a temperature less than 70°F. *The Commodity Storage Manual* (1995) says that the walnut in a shell kept at 0 to 7.2°C will last one year. A nut in the shell at 0 to –17.8°C will last two years. Out of the shell, a nut at 0°C will store for one year, it will last for two years at a temperature of –17°C. We have tried storing nuts in a fridge for several months, but by the spring we think that they taste rancid and the kernels rot. The Dordogne method of storage is to put the nuts in the eves of the house. In the English context and in the average house, freezing is better.

On the matter of storage, it is important to return to the subject of rodents. On this occasion the predator isn't the squirrel but the rat or the mouse. The storage area should not have any water anywhere near it. Here is another modern innovation that is a long way from the storage theories of Walter Fox Allen. His ideas of deep burial may well not stop a determined rodent. The area must be clean and the storage containers should be rodent proof. Plastic is good, but the vicious teeth can penetrate it if the rodent is determined. Also, plastic, if the outside atmosphere is warm, may make the contents sweat. Another suggestion is to hire a good rat and mouse hunting cat. When I asked the French about this problem their strong endorsement was for the cat. Alternatively using a trap on a regular basis will keep the rodents under control.

If you go into a supermarket, you will find new walnuts on the shelf. I was interested to see that they had been imported from the USA. The nuts were described as a variety called 'Hartley'. The label said that the nuts were harvested in 2008 and the sell by date was 2010. The nuts were vacuum packed but I suspect the advice was given with the understanding that they would be kept in a room temperature environment. It is possible to deep freeze walnuts so long as they are shelled and they are frozen for no more than a year.

Storage is most important and the right conditions do help preservation, although if the nuts are kept too long they will go off. An apple that is beginning to rot shows outward signs by its skin becoming crinkly. Inside the taste becomes less sour, the texture becomes rather woolly and even part of the inside may be going brown. Walnuts are dried and should not show any outward sign of deterioration. The evidence that they have declined is in the taste. It becomes rancid and bitter with a fleshier texture than a fresh nut. It is a taste that lingers and is unpleasant. Throwing the package away is the only solution.

In his *The London Art of Cookery* (1800), John Farley suggests that walnuts can be preserved in the following way:

> "Put a layer of sea sand at the bottom of a large jar. Then a layer of walnuts, then sand, then walnuts, then sand until the jar is full. Be sure the nuts don't touch each other. When wanted for use lay them in warm milk and water for an hour. Shift the water when cool and rub them dry. They will peel well and eat sweet."

One wonders where the clean sand came from, as the Thames would not have been that clean in 1800. The other consideration is how dry the sand has to be. Looking at modern ideas of how dry nuts should be, the answer is that it must be very dry.

14  Cracking the nut and preparing for sale

There are numerous and some obvious ways to crack nuts using gadgets which are commonly found. There are however some we do not see or hear about in the United Kingdom.

The use of a mallet, or the Trequot as it is known in France, seems a simple solution. Using this the method needs care so that the nut is damaged to the minimum. The complete half is the most valuable part of the nut because it is the part the confectioners really want. In many ways the mallet is both obvious and quick. It is the old method used in the Périgord. The Périgord cracker is a circular tray (shown below) which is about a foot wide and about two and a half inches deep. Naturally, the wood used is walnut. In the centre is carved an egg cup shape, about an inch and a half across which holds the nut to be cracked. There are two rims towards the outside of the tray. These hold the shelled nuts. It is estimated that two people will crack twenty-five kilos of nuts in a day. No doubt whilst working their hours a great deal of gossip will have been passed around. The tourist versions are nicely oiled and elegant enough to display on a dining table.

In the United Kingdom we have numerous tools in our kitchen cupboards all of which have differing capabilities.

On the left is a screw cracker. In the middle is a traditional Périgord cracker. The nut is placed in the cup in the centre. The shell is placed on the outside rim and the nut on the inside. On the right is the lever nut cracker commonly used in the UK.

Some alternative ways to crack nuts

An American idea is a simple one. Just drive your car over the nuts. There is little or no effort on the part of the person as the car does the work. I haven't tried it but, I think that it would be destructive and pulling the nut meat from the shell would take time.

In Japan the crow, an ingenious crafty bird, has found a clever way to break walnuts. The method was shown on BBC2 on 3rd January 2012 in a program called Natures Wierdest Events. The crow collects the nuts and transports them to a pedestrian crossing with a traffic light. The nut is placed on the road when the lights are red. The accurate crow will have the nuts crushed as the cars go over them. Another fairly wild idea is to put the nuts into a concrete mixer, but then there are a happy few hours sifting the nut from the meat.

None of these ideas are mechanized to give efficient productivity in this modern world of agriculture. The picture below shows a French machine to crack nuts in quantity.

There is not always an absolute need to take out whole nuts. Breadmaking needs crushed nuts and there are other recipes where the walnut needs to be crushed or small pieces are required. Salads do not need complete halves and for cake decoration halves may not be necessary.

In commercial terms the most valuable nut is the half. Cake makers will pay more for them than for chopped nuts.

This machine will shell 70kg of walnuts an hour or the equivalent of a day's work with a mallet in twenty-five minutes. The hopper is filled with nuts, and cogged rollers are set at a distance to allow the nut to pass but not be crushed on the way through. Not all the halves will be preserved perfectly and the kernel has to be sieved from the shell.

15  Walnut oil

The manufacture of oil from the walnut is an important rural industry for the Périgord area of France. It takes five kilograms of whole nuts to create two kilograms of kernels which, when crushed, will produce one litre of oil. The oil is made over the winter months between November and April. Essentially, the process has remained the same for centuries.

The kernels are crushed by a revolving vertical stone wheel for a period of forty minutes. This process creates a pulp which looks like a glob of pâté. It is then heated to a temperature of 60°C before being pressed for fifteen minutes to extract the oil.

The French make four types of oil:

Huile Vierge is cold pressed. The process is the same as has been described but there is no heat applied. This oil is tasty, but it does not keep very well.

Huile Concassée is oil produced for cooking.

Huile Blanche is ordinary oil.

Huile Noire is oil produced by dividing and reheating the shell.

Many people believe that when walnut oil is heated the flavour is destroyed and that it is best used with cold foods.

The process

I did some research on the way one oil producer works.

A mill wheel is used to crush the kernels. The wheel weighs 350 kilos. The flipper on the left-hand side pushes the kernels under the wheel as it turns. When the wheel starts, it will revolve round the spindle. As the kernels are crushed to a pulp so the wheel seems to skid over the top.

Next, the pulp is transferred to a heating pan over a wood fire.

The heating or cooking goes on for about 30 minutes and the temperature should be 80 degrees centigrade.

Mill used to crush the kernels.

Heating pan and mechanical stirrer.

The paste is shovelled out of the pan and placed in the press. The press is lined with jute sacking and the top is covered.

Jute sacking used to line the inside of the walnut press.

On top of the jute sacking is placed a block of oak called a Bonbon and the press is then pushed down on to the block. Oil immediately comes out from under the press.

Bonbon.

The press.

The process of making oil has not changed much for several centuries. Electricity made probably the biggest change in this important industry in the Dordogne.

There is a mill called the Moulin de la Tour where all the machines similar to those shown above are driven by a water wheel. The mill dates back to 1772, it is thought. It originally belonged to Lady Catherine de Monzie. She was the widow of

Joseph de Fenis, a King's Counsellor, Private Lieutenant of Présidial and Seneschal of Sarlat.

The mill had two functions: creating oil and milling wheat. Sadly, the wheat milling has ended, with the demand for bread increasing and the need for more power than can be provided from a water mill.

The water is stored behind the mill in a small reservoir fed by springs to provide the power.

The water wheel.

All of the machines in the mill are powered by the water wheel.

Walnut kernels being ground under a stone.

There have been three generations of millers at the Moulin de la Tour. Here the heating process has finished and the paste is about to be transferred to the press.

The press operating on the Bonbon.

The millers say that 30 kilograms of walnuts makes 15 litres of oil. As a comparison, the same quantity of hazels or almonds will make only 10 litres of oil.

Inside the jute wrapper is the residue after the pressing. This is called the tourtoc. After it has been used, it is not wasted because it is fed to farm animals including horses and chickens.

Storage

The storage should be at about 1 to 4°C. Humidity should be 60/70% in a dark or shady place.

Air must be kept out. In France the use of a floating lid in the container helps to achieve this exclusion.

16  Walnuts as a food

'And I dream about walnuts', shouted the monkey. 'A walnut fresh from the tree is so scrumptious-galumptious, so flavours-savoury, so sweet to eat that it makes me all wobbly just thinking about it.'

Quoted with permission. Roald Dahl: *The Giraffe and the Pelly and Me*, published by Jonathan Cape Ltd and Penguin Books Ltd.

In folklore the walnut was considered a food of the Gods. The word *Juglans* derives from "Jovis Glans" which means nut of Jupiter.

In earlier times in Persia, only royalty could eat walnuts. Other people were not allowed the privilege. Time has passed and thankfully, the rest of us can now freely eat them! The oldest archaeological site where walnuts have been found is in Iraq at the Shanidar caves. But the nut migrated and, just to prove how well the walnut has travelled, there was evidence of walnuts found in a Mesolithic dung heap in Switzerland. Walnuts were a tradable asset in the 10th century when farmers used them to pay rent. Obviously nutritious, a chemical analysis reveals valuable elements for a person's diet:

0.34 mg Thiamine
0.54 mg Vitamin B6
3.4 mg Manganese
1.6 mg Copper
158 mg Magnesium
346 mg Phosphorous
441 mg Potassium
3.1 mg Zinc
2.8 mg Iron
98 mg Calcium

It is said that the walnut has 69% fat of which 49% is polyunsaturated. It contains 6% Omega 3 which is an important chemical for a healthy heart. There is no cholesterol in a walnut.

Nut allergy

Apart from the nutritious value of the walnut, there has to be a mention of nut allergy before further food value can be discussed. A nut allergy is much more common than many people might think. There are many different symptoms of nut allergy. The face can swell and even more dangerous, the airways can block from the swelling. Anaphylaxis is life threatening. Low blood pressure, rashes and airway swelling are other symptoms.

The international trade in walnuts

Inevitably the walnuts we find in United Kingdom shops are imported because we don't produce any quantity here for our consumption needs. America is the largest exporter of nuts, mostly produced from small family farms. The United States exports 35% to 40% of the world production. The next largest exporters are Mexico, Moldova, China and France. China is the largest walnut producer in the world, but does not export as much as the US. China has a large population and (I assume) the people use the nuts as part of their daily diet.

If we listen to the food experts in books, the television or in the press, taste and ingredients are the making of any dish. The suggestion we hear time and again is that taste from too many ingredients becomes confused and that a concentration on a few specific tastes is the key to a good recipe. Walnut has a very particular taste which some people might say is slightly bitter. Like many nuts it has a rather dry taste and the outside seems quite fleshy. That taste can influence the ingredients of a dish, but it is also the case that walnuts can taste different. Some walnuts will be sweeter than others. David Sexton writing in the *Evening Standard* on 31/01/2018 extols the taste of French walnuts. He describes them as tasty in comparison to nuts produced in California or China. It could be the variety of trees or the soil in which the nuts are grown.

Food thoughts

I think the older nuts which may have been stored for several months, will have a rather sour taste, albeit they can be improved by soaking them in milk or in salt. The walnut is probably not as popular as other nuts like the almond or the peanut or cashew. However, the walnut retains a famous name within the confectionary business and amongst salad recipes. Its unique taste does not go together very well with every ingredient. The script writer in the film *King Kong* writes that the actor Adrian Brooke was offered "Lambs Brains in a Walnut sauce". Later in the film, there is a suggestion of an offer of "porridge with walnuts". Neither of these combinations seem suitable, but taste is a personal matter.

One way to consider recipes for the walnut is at the beginning of the day in the morning and at breakfast. For a healthy food person, breakfast could consist of yoghurt sprinkled with walnuts and a generous dollop of maple syrup or honey. At any other time of the day, walnuts will go very well with cheese. On the Fortnum and Mason walnut wafers packet, there is a caption that says "blue veined cheeses demand blue-blooded biscuits. Walnuts work winsomely with all colours of cheese." It is known that in the 16th century walnuts were eaten with stilton and port. (Vegetarians in Paradise website)

Mark Hix suggested in a *Sunday Times* magazine that using crushed walnuts as croutons for certain types of soup enhances the taste. He thinks this idea would go well with onion soup or butternut soup. Walnuts would be delicious with the famous Russian borsch because walnuts go so well with beetroot.

Walnuts in salads

A salad is one of the best dishes for the walnut. There are many types of salad in which the walnut can enhance the taste. The nut can be crushed or included as halves.

It is quite common to find walnuts in salads overseas. I paid a visit to Malta and went to the wonderful Palazzo Parisio in Naxxar. Their salad was a simple dish. The smoked chicken and walnut salad with olive oil was a perfect combination for a starter in a wonderful garden surrounded by high walls covered in Bougainvillea on a warm summer day. I am quite certain that this was not something the hotel invented, but we enjoyed our lunch and went back for a second lunch another day.

I believe that to enhance the taste of the walnut in a salad one should fry them quickly in butter and then set aside on greaseproof paper to cool.

Cheese and Walnut Salad

An alternative is a cheese and walnut salad. Goat's cheese, chopped walnuts and lettuce. It makes a really soft cheese salad.

Pear, Walnut and Roquefort Salad

A bed of lettuce is laid on the plate. The pears should be sliced very thinly longways, using about a half of the pear for one plate. The walnuts are halved and can be lightly roasted in butter and sprinkled into the salad. Roquefort cheese is diced and added to the salad.

Waldorf Salad

The salad is made with two or three chopped celery sticks cut into cubes, a small handful of chopped walnuts, two apples peeled and cubed, one lettuce, the juice of one lemon, salt and pepper to taste and five or six table spoons of mayonnaise.

Smoked Goose Breast on a Waldorf Salad with Garlicky Walnuts

The Rannoch Smokery in Kenmore, Perthshire suggests an addition to the original Waldorf salad recipe. Their method adds two tablespoons of oil, one tablespoon of chopped walnuts, two cloves of garlic, one tablespoon of chopped parsley and one smoked goose breast which has been sliced.

Having made the salad, heat the oil in a small pan, add the walnuts and garlic. The ingredients need to be cooked for two to three minutes until the nuts are golden. The nuts are allowed to cool for a few minutes before the chopped parsley is mixed in. A spoonful of salad sprinkled with the oil and topped with the goose breast will serve two people.

Scallop, Sea Bass, Orange Mayonnaise and Walnut Salad
Lightly sear seasoned scallops and a small portion of sea bass. Place the scallops and sea bass on a bed of salad and serve with the orange mayonnaise.

Game, chicken and the walnut
Game Salad
As a suggestion, this salad is possibly a summer dish when the deep freeze needs to be emptied.

Cut bread into croûtons depending on numbers. Take the meat from a woodcock, duck or pigeon and cut into chunks. Cut up a clove of garlic and leave to one side.

Melt a knob of butter in a frying pan and add a splash of olive oil. Add the game meat and fry over a medium heat, turn and season. Remove from the pan and let it cool.

Next fry the croûtons, add the garlic and thyme. Once the croûtons are crispy, remove from heat and add a handful of walnuts.

Mix a teaspoon of Dijon Mustard and some balsamic vinegar in a mug. By adding some olive oil the dressing is thickened.

On a plate of lettuce add the game, croûtons and the dressing.

Pheasant and Walnut Terrine
This is a terrine invented by Francis Binney with a little help from me. Take two skinned pheasants. Remove the breasts and the legs.

Mince the breasts and livers with pork fat.
1 glass of Madeira, Brandy or Calvados to taste.
1 finely chopped onion.
200g butter.
200ml cream.
200g breadcrumbs.
250g walnuts, chopped, but not too finely.
500g streaky smoked bacon.
20g green peppercorns.

6 bay leaves.
Several juniper berries (if liked).
Salt and pepper.

Butter a loaf tin in which the terrine is to be made. Bay leaves, juniper berries and green peppercorns are placed in the bottom of the tin. Line the bottom and sides of the tin with the rashers of bacon, leaving the ends hanging over the edges. Place half of the mince in the bottom of the tin. Add a layer of pheasant breast. On top of the breasts place the remainder of the mince and then cover the top by folding over the bacon. Cover the tin in a double layer of greaseproof paper. Cook in a bain-marie for one and a half hours until the terrine pulls away easily from the tin. Place the tin on a plate and then press under a weight for six hours and refrigerate or freeze.

An alternative is to use pickled walnuts in place of the dry walnuts. This will change the taste, but it will still be delicious.

Pheasant and walnut terrine in a festive setting.

Walnut Sauce for Pheasant, Chicken or Fish

450ml of walnut oil.
200ml boiling water.
5 thinly chopped cloves of garlic.
2 tablespoons vinegar.
Half a tablespoon of salt.

To give the sauce more taste add one tablespoon of ground saffron, three tablespoons of ground coriander and a quarter of a teaspoon of paprika. To finish off the sauce add a small amount of cayenne pepper.

Amy Willcock describes a Pheasant with Walnuts and Madeira

1 jointed pheasant.
1 large onion chopped roughly.
2 rashers of bacon.
1½ tablespoons butter.
25g walnut halves.
275–300ml game or chicken stock.
1 tablespoon redcurrant jelly.
Salt and pepper.
1 tablespoon chopped parsley.

Use a large casserole dish. Brown the joints in butter and then remove to a plate. Put the rashers of bacon and onion in the dish. Cook until lightly browned. Add the Madeira and stock to cover the joints and then bring to the boil. Add the redcurrant jelly. Reduce to simmer for one and half hours. Serve with a swirl of redcurrant jelly.

As an alternative, place two one-hour browned pheasants in an orange juice and Madeira stock. In the meantime, boil the orange rind in water until it is soft, drain and rinse it. Melt a small knob of butter in a frying pan, add a few shallots, bacon lardons and crushed walnuts to taste. Joint the pheasants and put in a small serving dish. Thicken the juices with cornflour and add the frying pan contents and rind. Add salt and pepper to taste.

Elisabeth Luard described in the *Country Life* issue of December 15th 2010 the preparation and cooking of a "Circassian pheasant"
The Georgians cook their pheasants in a particular way. The pheasant originated in the Caucasus and they are prized by the Ottoman Turks. This recipe feeds four people. You need two pheasants, one onion quartered, one carrot roughly chopped, a few parsley stalks, salt and pepper.

To go with the pheasant there is a sauce. You need 170 grams of freshly shelled walnuts, 85 grams of white breadcrumbs, one clove of garlic skinned and roughly chopped, one teaspoon of ground cinnamon, salt and pepper. To finish salt and pepper one to two tablespoons of walnut oil and half a tablespoon of chilli. The cooking method, Elisabeth says, is to wipe the pheasants and remove surplus feathers. Place them in a pan with herbs salt and peppercorns and enough water to cover them. Boil and then turn down the heat. Place the lid loosely and leave to simmer for about an hour and a half. Remove the birds when tender, skin and debone then cut the meat to bite size chunks. Remove the bones to the broth and bubble up until reduced to about a pint. Strain the broth and reserve.

Dry roast the walnuts in a heavy pan for a few moments until lightly roasted. Do not let them burn. Drop the nuts and the other sauce ingredients into a food processor and add a ladle full of broth. The process turns the mixture into a thick purée. Thin it and return it to the remaining broth. Turn the pheasant pieces in the broth and then pile onto a dish. Trickle the oil into the broth and add the chilli.

Fesenjan

Fesenjan is an Iranian dish for special occasions. (*Elise Bauer Simple Recipes Food and Cooking Blog* 2012). This dish is a chicken stew with a pomegranate molasses and walnut sauce. The pomegranate molasses is made from pomegranate, sugar and lemon juice.

The walnuts are roasted for 8–10 minutes at a temperature of 180°C. They are cooled and blended to a finely chopped state.

Cut 1–1.5kg boneless, skinless chicken joints into pieces. Melt one tablespoon of butter and one tablespoon of olive oil in a pan. Place the chicken pieces in the pan and cook until bronze-coloured.

Using one tablespoon of butter and one tablespoon of oil to sauté two chopped onions. Return the chicken pieces to the pan, add 500ml of chicken stock and simmer for 30 minutes.

Add one tablespoon of salt and sugar to taste, one teaspoon of pepper and one and a quarter teaspoons of saffron. Then mix in the ground walnuts and pomegranate molasses. Simmer for one hour.

This dish should be served with rice.

Turkey Stuffing

Rick Stein has a stuffing for turkey as follows:

    30g of butter
    1 large onion finely chopped
    7g pancetta very finely chopped
    2 tablespoons chopped fresh thyme
    4 tablespoons chopped fresh parsley
    Finely grated zest of 1 lemon
    2–3 anchovy fillets finely chopped
    50g walnuts chopped
    150g fresh breadcrumbs
    1 large free range egg to bind

Melt the butter in a frying pan over a medium heat. Add the onion and cook for five minutes to soften, then stir through the chopped pancetta and cook for five minutes more. Remove from heat, then add the remaining ingredients. Season and then allow to cool before using to stuff the neck of the turkey.

The beetroot and walnut

The first and most lasting impression of the beetroot is the deep red colour. The taste is unique, a bit earthy and sweet.

The beetroot crop is best harvested in September or October. The plant is not difficult to grow, but needs a fine tilth in the seedbed. The seeds are small and the seedling needs to find the light. Water and fertilizer must be added and a neutral soil without excessive acid or alkaline properties. In clay areas some lime will sort out the acid alkaline balance. Most seed packets advise that seedlings should be thinned and this allows better growth and bigger beetroots.

I like the beetroot taste, but it can be improved. Most of the recipes marry beetroot with other ingredients. It is commonly thought amongst chefs that the walnut has a taste that goes well with beetroot. The tastes complement each other well.

Warm beetroot with cream or cream cheese, Parmesan and a sprinkling of walnuts goes well with the comfort food of a jacket potato with butter and pepper and a cold meat. The beef, lamb, pork or poultry could be left over from the Sunday roast. Alternatively, smoked breast of duck can be used as this strong flavour would be nicely complemented by the beetroot, Parmesan and walnuts.

Walnuts warmed in butter, then spread over cooked beetroot.

A beetroot salad can be served with mustard, horseradish and a soft-boiled egg.

I was asked to a drinks party. Our hostess had diced a beetroot into one-centimetre cubes. She had made a thin puff pastry and formed it into a pie shape three to four centimetres across. The miniature pie was filled with cream cheese, and she had placed a few pieces of diced beetroot on the cheese and then sprinkled small walnut pieces on the top. This made an excellent and tasty canapé.

Another very simple idea for a canapé is to almost slice a date in half leaving the bottom skin intact. Then place a small dab of blue cheese in the date with a walnut half.

Spreads and dips
Carley's Walnut Butter
This butter is made in Cornwall by the Carley family company. They buy their supply of walnuts from Moldova. The nuts are grown under Soil Association conditions and arrive shelled, clean and with a moisture of 6%. If the supply runs out the company turns to India. They prefer the Moldovian nuts because they have a sweeter taste.

The company makes about 12,000 jars from two tonnes of walnuts. They are sold to health shops particularly because of the presence of Omega 3 in the nuts. The recommendation is to use it as a spread, in sauces, dressings, dips and baking according to the label.

Honey
I bought a pot of walnut honey. It was made in France by Bernard Porlefaux and was called Miel de Châtaignes au Noix (chestnut honey with walnuts). The honey was clear with half walnuts "floating" in it. The taste was slightly musty and the honey was a bit bitter with the walnuts. The question in my mind was whether a different type of honey might have improved the taste?

We tried the honey and walnut with some wild duck and asparagus. It was impressive but slightly sweet. The dish needed something that would take some of the sweetness away. My suggestion would be to add lemon or lime.

The Honey Association says that flavour in honey "closely mimics the characteristics of the herb or tree whose flower the bee has visited". These flowers could be orange blossom, lime blossom, rosemary or thyme. The only flower that doesn't grow in this country is orange, but the beekeepers in the UK are harvesting honey from apple blossom, hawthorn, lime, dandelion, borage, heather and rape.

Cabechu on Croûton with a drizzle of Honey and Walnut Bits
This is a recipe from France.

Cabechu is a goat's cheese local to Rocamadour in France. This is a continental Welsh Rarebit in the way it is cooked. The croûton is toasted and the cheese grilled. A small quantity of runny honey is spooned over the cheese and sprinkled with small bits of walnut.

You can put this dish onto a bed of salad with walnut oil to taste.

Mahammra
This is a recipe from the Levantine (*epicurious.com*) for a dip to serve with crudités or on pitta bread. As a dip at a cocktail party, it might be worth a try.

175–265g roasted red peppers drained.
60g breadcrumbs.

40g walnuts finely chopped.

2–3 cloves of garlic minced.

1 tablespoon lemon juice.

1 tablespoon pomegranate molasses.

½ teaspoon ground cumin.

½ teaspoon chilli flakes.

2–3 tablespoons extra virgin olive oil.

½ tablespoon salt.

Blend all the ingredients except the olive oil in a blender or food processor until smooth. Stir in the oil gradually. Transfer to serving dish and serve with parsley.

Cakes and confectionery

Walnuts have been used in confectionery and cakes for a very long time. The producer will probably aim for this market as perfect half walnuts are prized by confectioners and cake makers and thus command the best price. Apart from the taste, the walnut half will act as a very acceptable decoration. One of my favourites is the coffee and walnut cake.

Coffee and Walnut Cake

115g butter or margarine.

2 eggs.

115g self raising flour.

115g caster sugar.

2 tablespoons of instant coffee mixed with warm water.

Crushed walnuts to taste.

Bake in an Aga for 45 minutes or in an oven at 180°C. The two layers of cake are topped with coffee flavour butter icing.

The Fuller's Walnut Cake

The Fuller's teashops were started in 1910. There were 24 shops at the height of the business. On the menu featured fish and chips and other traditional British food, but they were famous for their walnut cake. A lady I spoke to said she had frequently visited a Fuller's teashop in Liverpool as a child for a treat. I asked her to describe the cake and to compare it with the delicious coffee and walnut cake we were enjoying at the time. The cake we were looking at was in two layers with butter icing and walnut halves on the top. The Fuller cake, she told me, was in three layers and fairly similar in taste. Sadly the Fuller company went into liquidation in 1974.

The Walnut Whip

The Nestlé Walnut Whip is world famous and has been manufactured since 1910. It is their oldest product, but it was originally made by Duncan's of Edinburgh. The whip has a distinctive cone shape and is made these days from milk chocolate and soya lecithin. The soya is used as an emulsifier to which is added butterfat, flavouring, sugar, walnuts, glucose syrup, glucose fructose syrup, dried egg whites, humectant, flavouring and tartaric acid. The walnut has been placed on the top of the cone since its invention and is no doubt the reason why the Walnut Whip got its name. For a while there was also a walnut inside at the base of the cone. The second walnut was removed in 1966, and in 2017 the makers decided to exclude the walnut on the top of some Whips, explaining that the cost of walnuts had risen!

Puddings and desserts

The walnut is sweet and is an obvious ingredient to add to any pudding. As we have seen from the confectionery descriptions, the walnut is a topping either as halves or crushed. The nut complements chocolate and coffee. Both these ingredients are often seen in puddings.

In France, the crêpe is a famous pudding. In Britain we call it a pancake, and it is made from flour, eggs and milk. Historically, crêpes originally came from Brittany.

A chocolate covered crêpe with walnut chips is an easily-made dessert, which is both filling and delicious. Crushed walnuts can also be sprinkled over other desserts such as ice cream or even tarts with cream.

This crêpe was produced at the Belvedere Restaurant in Domme, Dordogne. Crushed walnuts are sprinkled on top.

This is a dish produced in Argentina. A simple tart with slices of apple accompanied by a sprinkling of chopped walnuts and a scoop of ice cream.

Baclava

Baclava is Middle Eastern dish. It is five layers of filo pastry with walnut syrup between each layer. The following ingredients are needed.

> 225g shelled walnuts
> 50g soft brown sugar
> 2.5 ml ground cinnamon
> 450g filo pastry
> 150g melted butter
> 175g clear honey

The method of cooking is to grease a 24 by 18cm roasting tin. Mix the walnut, cinnamon and sugar in a bowl. Halve each sheet of pastry to measure a 25cm square. Fit one sheet of pastry into the bottom of the pan and fold the surplus up the sides. Brush with melted butter. Sprinkle one fifth of the nut mix over the top. Repeat the process four times. When the fourth layer has been put in place put the rest of the pastry on the top. Add a squared pattern to the top of the pastry.

Bake in the oven for 15 minutes at 220°C and then for a further 10–15 minutes at 180°C until the mixture is golden brown. Meanwhile warm the honey over a low heat in a pan, pour over the baclava and then allow to cool for an hour and a half.

Traditional Georgian Walnut Cake

At the Tamada Georgian restaurant in St John's Wood in London, Idacali is offered as a dessert. This is a traditional Georgian walnut cake. Honey and flour are mixed to create a wafer thin layer. This is baked and, when cool, ground walnuts and cream are spread over the honey and flour wafer. The process is repeated six to eight times. The cake is then cut into a diamond shape. This cake is sweet but delicious.

Walnut Ice Cream

The Galleria in Milan has several fine restaurants, providing some excellent regional food. Savini is one of the finest and to one side of the restaurant is a shop that sells delicious ice cream. One of the ice creams is a walnut vanilla. It has a brown and white look about it. It is not a gritty nutty ice cream. It is smooth with a clear flavour of walnut.

Walnut Maple Crème Brûlée

Heat oven to 140°C.

Six ramekin dishes are needed with a large roasting tin.

Heat 500ml of cream over a medium heat in a saucepan adding 100ml of maple syrup. In the meantime, split a vanilla pod from end to end and then use a knife to scrape the seeds into the cream. When the cream begins to bubble along the edge of the pan, slowly pour the cream into a six egg yolk mixture, stirring constantly, and add the remaining cream into the mixture. The next stage is to sieve the cream into a jug. You now have a custard-like mixture. You can crush walnut halves and add to this mixture, but this would be to taste. Pour the custard mixture into the ramekins and place them in the roasting tin. Add hot water to the tin, so it comes halfway up the side of the ramekins and then bake for about 45 minutes. Insert a skewer into the custard and if it comes out clean, the brûlée is cooked.

Take 12 walnut halves and place them on a baking tray and bake in an oven for about ten minutes.

Using a tablespoon of maple syrup, we make the topping. The aim is to form a sugar disc by using a blowtorch. The alternative is to use granulated or superfine brown sugar. Place a tablespoon of sugar on the custard mix and then use the blowtorch to caramelize the sugar.

The mixture should then be placed in the fridge for some hours or up to two days.

Charoset

The traditional apple and walnut Charoset is a Hebrew dish served at Passover. The Mishna describes the foods to be placed on the Passover table. It is a Jewish tradition.

3 medium Gala or Fuji apples, peeled, cored, and finely diced
190g cups walnut halves, lightly toasted, cooled, and coarsely chopped
150ml sweet red wine such as Manischewitz Extra Heavy Malaga
1½ teaspoons ground cinnamon
1 packed tablespoon brown sugar

The ingredients need no cooking. They are mixed into a bowl stored at room temperature until ready to be served. (*epicurious.com*)

Walnut and Coffee Tart

This is a variation on an American favourite – pecan pie.

Line a 20cm loose-bottomed tart tin with sweet dessert pastry. Prick the base and line the pastry with foil and baking beans. Bake at 190°C for 15 minutes. Remove the beans and foil and bake for a further 10 minutes until golden brown. Allow to cool.

Filling
1 tablespoon of instant coffee
175 ml maple syrup
25g butter
175g soft brown sugar
3 medium eggs, beaten
1 teaspoon vanilla extract

Heat the maple syrup and coffee in a small pan until nearly boiling, then allow to cool.

Beat the sugar and butter together, then beat in the eggs gradually. Mix in the maple syrup and vanilla extract.

Arrange walnut halves (about 110g) in a single layer in the pastry case, then slowly pour over the filling. Bake at 180°C for approximately 30 minutes until firm and lightly browned.

Allow to cool slighly, then remove from tin. Serve warm or cold with cream or ice cream.

Bread

Many nuts are added as an ingredient to bread. The walnut taste is well suited to a bread mix and has health benefits.

Panepato

Panepato means pepper bread in Italy. It is a mixture of pepper, cinnamon, nutmeg, corinth, raisins, crocus, coriander seed, caraway seed, fennel seed, fresh grapes, dried figs, almonds, hazelnuts, pistachios and walnuts. It is called Pane Impepato in Siena. The people of Florence call it pandigust which is a flavoured bread. This is made from selective ingredients, mixed into a batter and boiled refined sugar and trimmed with marzipan paste, worked in various manners and glazed sugar. There was an inferior grade of bread that did not have the select ingredients. This version was limited to simple batter made from poorer grade flour made with wheat containing shapps, pepper, walnut, dried figs and honey. This description is taken from Giovanna Giusti Galardi's book *Sweets at Court, Paintings and Other Tasty Treats*.

Walnuts and wine

Many wine merchants say that nuts of any sort will destroy the palate of a drinker of fine wines. There is a dryness in the taste of a nut which lingers. Some nuts leave more dryness than others. Walnuts do have a dry taste, but I was able to get opinions from some merchants about which wines could be drunk with walnut recipes.

Thinking of salads, one wine merchant suggested a dry red wine, possibly a Burgundy.

Another suggested that with candied walnuts an Italian sweet wine called Vin Santo Fontodi. This wine is sometimes used as Vino Sacro. It is made from grapes grown in the Chianti district of Italy and when picked they are dried on straw mats. The process concentrates the sugar.

The writer and author Victoria Moore declared in the *Daily Telegraph* that she had a fondness for sherry. She suggested several types of sherry with jamón ibérico. She suggested sherry with tapas, tempura, croquetas and calamari. She mentioned sherry with Scotch eggs or veal chops, pheasant partridge and pigeon. There are some ideas in here that strike a note and I think provide good advice. Victoria mentions that a few walnuts with a glass of sherry after work is a good early evening aperitif.

David Sexton writing about walnuts and wine in the *Evening Standard* on 31/01/2018 suggested fresh French walnuts with Maury. Maury is made from the Grenache grape. The wine is made either as red, rosé or white. The vines are grown on the steep limestone hillsides on the eastern side of the Pyrenees in the Languedoc region of south west France.

17  The autumn and winter

The nuts have been picked and the days begin to shorten. The grass and weeds stop growing and the work in the groves draws to a close for the winter months. Within the tree the sap stops rising and any water supply to the leaves is cut off. This causes the leaves to turn to their autumn brown colour and then fall.

During the winter the tree is asleep to all intents and purposes. There are not many events that will affect it. The buds are immature and the frost does not seem to hurt them. Any sap left in the tree may freeze, but this does not produce long-term damage. The roots under ground are unlikely to be affected by frost. The cold would have to go down a long way. We do not have prolonged cold periods on a regular basis (1963 was an exceptional year when snow was on the ground from Boxing Day to March).

This walnut tree that has lost almost all its leaves on a cold morning following an early frost. The picture also shows the growth the tree put on in the current year and next year's bud is prominent at the end of the twig.

The biggest risk is wind. An exposed site could see trees blown over or branches being snapped off. An immature tree blown over is worth little or nothing and a branch snapped means that there will be bleeding which will damage the tree and effect the harvest in the subsequent year. Yet, as we have seen earlier, a successful grove will have to be in an area where there is wind so that pollination can take place. Unlike other trees, the walnut does not impress with the finest autumn colours. There are no wonderful reds or maroon colours.

The shell

At the end of the harvest and in the autumn period there is the shell.

The shell is like a bone. It is hard and wooden albeit smoothish to the touch. It has been described as the ovary of the nut. It would seem that when the shell has been broken, there could be no further use for it, but there is.

The barber to Louis XI used the heated edge of a walnut shell to shave his master. I dread to think how uncomfortable this must have been. Walnut shells are abrasive and have an ability to clean. The shells are crushed up to either a fine or coarse consistency. They are non-toxic and in time they will biodegrade. Some bakers spread walnut shells over their ovens to prevent the dough sticking to the sides. The method is to crush the shell to a powder.

Apparently walnut shells are used to clean aircraft engines. That is to say jet engines and helicopters. According to an ex-naval engineer, the Sea King helicopter, now out of commission, had two Rolls Royce Gnome engines. Engineers clean them from time to time. The substance used is known as Grit Lignocellulose Turco softblast 45/60 grade CS59. The engine runs slowly at idle and about a pound of the Turco softblast is sprayed into the compressor which has a multi bladed disc at the front. As the Turco is shot through the engine, the material is soft enough not to damage the engine itself and has the effect of sponging up soot from the lining of the engine. As the Turco goes through the combustion chamber, it burns and comes out of the exhaust as smoke leaving no residue. The engineers will wash the engine but say that the engine improves dramatically after this cleaning. It is said that a jet engine needs about 90 kilos of walnut Turco grit.

It may be one of those questions which few people would ever think of asking, but how do brass musical instrument makers create the shine? In days gone by, they would place the trumpet in a barrel with walnut shells and slowly revolve it to shine the surface. It is not done this way today. Another method used is to revolve the barrel with maize.

The shells are also dust free. This is an important asset to the pet industry. The shells are used to provide bedding for reptiles and birds that when in captivity need careful husbandry. Cleanliness and a dust free atmosphere are important for their wellbeing. Because of the nature of the shell, faeces and urine will be absorbed and once taken out the bedding will biodegrade and may have a benefit as a fertilizer.

In the Dordogne we saw the shells being used as mulch under roses at an abbey.

With the quantity of shells they have at the end of the autumn, it would seem that this is an obvious way to keep the weeds down and keep the moisture in the ground in the hot sun. It is possible that the rather abrasive nature of the nut might put off slugs if they are around.

The shells are used for other practical and artistic purposes. The town of Nontron in the Périgord is a centre for the making of cutlery, in particular knives for very many uses. There are penknives, key-ring knives, fish knives with back fins carved in the wood as well as knives for the table. The knives have been manufactured in the town since 1653. The hilts of these knives are made principally of boxwood and inscribed by burning with a special but unknown insignia. Quite remarkably, the company creates miniatures of their knives which are packaged in a walnut shell. They take the two halves of the shell, put a brass-hinged collar around the edge with a small catch so that the shell can be opened to reveal its miniature contents. Seeing these walnut shells and knives made in such a masterful fashion reminded me of the miniature furniture made by apprentices in the 17th and 18th century. These pieces are more of a collector's item rather than any practical use. Yet for somebody who likes the strange but rare artefact in a display cabinet, this may be very fitting.

Miniature knives kept in a walnut shell.

A use for the walnut shell was described in the Rudyard Kipling novel *Kim*.

> "My husbands are also out gathering wood." She drew a handful of walnuts from her bosom, splitting one neatly, and began to eat. Kim affected blank ignorance.
>
> "Dost thou not know the meaning of the walnut-priest?" She said coyly, and handed him the half-shells.
>
> "Well thought of." He slipped the piece of paper between them quickly.
>
> "Haste thou a little wax to close them on this letter."

Kim later goes on to say the letter should go to Babu. This is a piece out of an early spy thriller where the walnut shell is used as a disguise for a secret message.

There may be some other ways the walnut shell halves can be used.

The romantic could split the nut and place a ring in the nut for his loved one amongst a bowl of walnuts. The problem here is that the real receiver must pick the right nut.

Walnut shells being a home for small things is a part of the story of Thumbelina when she is given a beautiful polished walnut shell in which to sleep. *Thumbelina* by Hans Christian Andersen (1835) is part of a series of seven fairy tale books. An English translation was published in 1846.

The half shells have also been associated with toy ships. Small masts stuck in the bottom of a half nut with Blu Tack or some other adhesive, with paper sails, shaped like Viking ship sails, have been used for decoration and indeed for games. In a small pool, a large glass bowl or baby bath they will float. If the water is disturbed, their seaworthiness is called into question, but the decoration is appealing in calm waters.

18  The health issue

The walnut looks like a brain and historically people believed that as it looked like a brain, it would have medical value to that most vital organ.

The great Pliny is said to have used the walnut to cauterize teeth. I have never heard of or met a dentist who has suggested this way to cure a bad tooth.

John Evelyn says that the water of the husks of a walnut is sovereign against pestilential infection and that the leaf is to heal inveterate ulcers.

The unripe juice is said to act as a laxative. I think it would need some will power to consume such a concoction. To back up this statement, it is said that if the juice is mixed with honey, it can be gargled for mouth sores. It can also be used for wounds, gangrene and carbuncles.

According to M. Robinson, "The husk and shell and peel of the walnut are sudorific (sweat inducing) and are used to expel tape worms." (*Trees of the British Isles in History and Legend*)

The nut and leaf are said to cure other ailments. Herbalists and homeopaths use them in their cures. The rind for example can be made into a tonic to relieve diarrhoea. The nut helps menstrual dysfunction and eczema. The inner bark of the tree is used as a tincture for constipation. The leaf too is used as a mild laxative and for acne. It can also be used to bring down swelling. The black walnut is used for the cure of athlete's foot.

According to M. Grieve's *Modern Herbal*, walnut leaves and bark can be used for skin conditions, laxatives and as an astringent. The book was written in 1931 as a modern book on medieval folklore, cultivation and scientific medical facts about a range of plants.

There is a recipe from the same book on how to preserve green walnuts in syrup:

> Take as many green walnuts as you please about the middle of July, try them all with a pin, if they go right through them they are fit for your purpose; lay them in water for nine days, washing and shifting them morning and night; then boil them in water until they be a little soft, lay them to drain; then

pierce them with a wooden skiver, and in the hole put a clove, and in some a bit of cinnamon, and in some candied lemon rind: then take the weight of your nuts in sugar, or a little more: make it into syrup in which you boil your nuts (skimming them) till they be tender; then put them in Gally pots and cover them close. When you lay them to drain wipe them with a course cloth to take off a thin grey skin. (They are Cordial and Stomachial from the family Physician by Geo Hartman, Phylo Chynist who liv'd and travell'd with Honourable Sir Kenelm Digby, in several parts of Europe the space of seven years till he died').

In the 17th century Nicholas Culpeper used walnut leaves, honey, onion extract and salt to drag out snake poison. As our only poisonous snake is the adder, one can only assume that this was the biter.

"[Ripe] Walnuts can reduce inflammation and regulate blood pressure due to notable potassium" according to Marber.

The Louisiana State University research found walnuts in the diet can bring beneficial changes to gut microbiota boosting health and immunity. (*The Times* 20/03/2018)

We all have to work and some of us find our way into an office. For most of the day, we sit in front of a computer screen. Working in this way makes us relatively inert and we are not taking much exercise. Doctors tell us that this is not a good lifestyle. Exercise is one thing, but diet is another matter. The eating of nuts is said by doctors to be "a natural health capsule". We should eat vegetables, drink water, keep off fatty foods and eat walnuts. The last ingredient is my own addition, but it is wise council based on the scientific evidence that walnuts have quantities of Omega 3. This Omega 3 is known to reduce cholesterol levels in the blood. It has been suggested that if five nuts or 30g are eaten a day, cholesterol levels will either fall or be maintained at low levels. Weight loss is also thought to be helped by nuts.

At a family Christmas we have the traditional turkey, bacon, roast potatoes, roast parsnips, sprouts, bread sauce and the other usual trimmings. We then have the Christmas pudding, mince pies and brandy butter. In order to round the meal off, I offer some homegrown walnuts, adding that they have this important health benefit. My doctor daughter remarked that it would be an interesting question to wonder how many walnuts it would take to offset the fat in the meal we had just had. The British Heart Foundation says that most people will eat a Christmas meal which has the equivalent saturated fat of a pat of lard and the same amount of salt as 50 packets of crisps. Who knows the effect walnuts might have in reducing the bad effects of the rest of the meal?

The World Health Organization recommends nuts as part of people's diet to reduce cardiovascular disease.

19

Art and the walnut

Leonardo da Vinci is said to have used "pressed walnut oil". Other artists who used the oil included Van Eyck, Van Dyke and Raphael.

Italian artists sometimes painted on walnut panels. Leonardo da Vinci painted on poplar board in Italy, presumably by choice, but he also painted on oak when he was in France. The world-famous *Mona Lisa* was painted in 1503 by Leonardo and can be seen at the Louvre in Paris. The lady in the picture is said to be Lisa Gherardini. There is a copy of the painting in the Prado in Madrid, thought to be the oldest copy in existence. The painting was found in a vault at the Prado and research was started in 2011. It was not painted on poplar, it was painted on walnut board and executed in Leonardo's workshop. Andrea Salai was a favourite pupil of Leonardo and the artist most likely to have copied the great painting. Salai was born in 1480 and was the son of Pietro di Giovanni, a tenant of Leonardo at his vineyard. Francesco Melzi was born into a wealthy Milanese family, his father was a military engineer and spent time in the Milanese court. As a young man he met Leonardo, became his pupil and learnt from his master how to paint. He could have painted the copy.

The copy is believed to have been painted at the same time as the original, with the copier following Leonardo's methods. Yet, the copy was changed. The eyebrows are different and there is colour in the dress in the copy. This is all very mysterious, but the walnut board is part of the legacy of the copy.

Tom Norton observes that walnut oil is not made from walnut husks as it was at the time of Leonardo and Rembrandt. The old walnut ink had acid in it and could cause deterioration of the paper. It would also fade to a yellow from its darker colour. Modern walnut ink is acid free, however.

Returning briefly to furniture, it is possible to make a stain for oak and other timbers by boiling walnuts in a cheesecloth which will take in water.

The water changes colour and the boiling continues until the required colour is obtained. The longer the boiling, the darker the colour. Fortunately, it is no longer necessary for a restorer or cabinetmaker to make their own stain. This can now be purchased as Van Dyke crystals. Crystals are made from walnut husks and while the crystals are boiled, the colour will be determined by the quantity of crystals added to the water.

In the magnificent cathedral in the heart of Chichester in West Sussex are some rare wall paintings by the artist Lambert Barnard. The painting was commissioned by Bishop Sherborne in 1534 and it was completed in 1536.

Bishop Sherborne was born in 1454. He was educated at Winchester College. In later life he became ambassador, secretary and councillor to Henry VII as well as Bishop of Chichester. He performed an important service for Henry VIII when he sought permission from Rome to allow Henry to marry Catherine of Aragon after Henry's elder brother Arthur had died.

The painting is about four by ten metres and is on four oak panels covered with canvas. Each panel is joined by a standard tongue and groove joint. The painting is in two halves. To the left is a kneeling Saint Wilfred, who started the cathedral in 1075 AD asking the once pagan King Cædwalla to be allowed to build the original cathedral which sadly was burnt down in 1187. The cathedral was rebuilt in 1187 and completed in 1199. On the right hand panel, Bishop Sherborne is depicted pleading with Henry VIII to confirm his church in Chichester as written in the picture. Henry grants his petition.

The picture contains propaganda some of which is not clearly understood today. There is a monkey chained to a pillar in the bottom right hand corner of the picture and at its feet is a collection of walnuts. Some of these nuts are whole and some are in perfect halves. Experts have different views as to why they are there and what their significance might be. Karen Coke, an art historian and expert on Lambert Barnard says, "Barnard's charter paintings carry several layers of meaning, cleverly interwoven, so some pictorial elements can stand for several quite different things and the monkey and its nuts come into this category".

There are some explanations. According to Huizinga writing in 1927 the walnut is said to signify Christ. The sweet kernel is his divine nature. The green and outer pulpy peel is his humanity and the wooden shell the cross. James Hall in his *Dictionary of Subjects and Symbols in Art* says "The symbolism of the nut, generally depicted as one half of a split walnut was elaborated by [Saint] Augustine. The outer green case was the flesh of Christ, the shell the wood of the cross and the kernel his divine nature."

Karen Coke says that the broken walnuts in the hands of the monkey are indicative of the pagan Cædwalla's discovery and acceptance of Christ.

The monkey has piercing eyes, an intense stare and possibly questioning look. Catherine of Aragon had a pet monkey that was chained and there is a ring near the monkey. Could the monkey have something to do with Catherine? Could the half nuts be a symbol of a broken marriage? These ideas are tentative with little evidence, but Sherborne may have seen Henry's divorce coming and this could be a

A monkey and walnuts. Detail of
Lambert Barnard's painting in Chichester
Cathedral.

warning. Karen Coke suggests the "the monkey as an animal who can never achieve
the status of humanity because of his base instincts is here as a symbol of barbarity
and by extension representative also of Cædwalla's barbarism / pagan state. Hence he
is depicted willingly chained to the column behind him. The column being representative of the faith of the church now accepted by Cædwalla." Sherborne was elderly
by the time the painting was planned. He would have known senior members of the
clergy. The mystery is what he was thinking when he planned the picture. What did
the folk of Chichester think of it and did they understand the propaganda?

Walnuts often appear in still life paintings in the 17th and 18th centuries. Luis
Meléndez was born in Naples which, in 1716, was a dominion of Spain. His father
was a founder of the Royal Academy in Madrid. Luis wanted to become a court
painter and applied on two separate occasions, failing both times. His career was
blighted by a dispute his father had with the Academy and he died in poverty. Yet in
the last 20 years of his life he painted 100 still-life paintings for which he is famous.
The walnut and orange still life painted in 1772 hangs in the National Gallery in
London and uses the typical Spanish vertical technique. The nuts in the bottom right
of the picture are on a plate surrounded by a group of oranges behind and some
taller jugs, a barrel and some boxes. Here again some of the nuts are open. There is

no suggestion that there is some inner meaning to this and it is probable that Meléndez wanted to show the viewer what the inside of the nut looked like.

Luis Meléndez: *Still Life with Oranges and Walnuts*, 1772, oil on canvas.

Jan van Huysum (1682–1749): *Fruit and Flowers*, 1720s, oil on panel.

Pieter Claesz: *Still Life with Roemer*, 1644, oil on panel.

Giacomo Ceruti (1698–1767): *Still Life*, 1750s, oil on canvas.

Willem van Aelst (1627–1683): *Still Life of Fruit*, *c.*1667, oil on canvas.

Jean-Siméon Chardin (1699–1779): *Still Life with Peaches, a Silver Goblet, Grapes and Walnuts, c.*1760, oil on canvas.

 **Miscellaneous**

I have often thought of trees as being the silent witnesses of history, not just whilst they are living but also after they have been transformed into furniture. The life of a walnut could be 300 years and the furniture, from which they are made, for many years thereafter. I think of the oak tree in which the future Charles II hid to avoid capture by the roundheads. Then there is the apple tree under which Adam and Eve stood whilst being tempted by the serpent. I would have to include the apple tree under which Isaac Newton sat when he worked out the theory of gravity as the apple fell to the ground.

More generally I can imagine the meetings people had, the romantic, the political intrigue, the battles, and more macabre, the executions from their branches.

Legends and myths of the walnut are not common in the UK, but there are some to be found in Europe. Legend has it that in Italy witches covens met under walnuts to weave their spells. Why they should pick the walnut is not clear.

Walnuts are also life giving to humans by their gift of nuts. In the Alps, two walnuts and two glasses of liqueur are placed in front of a bride and groom at their wedding. The two halves of the nut are of course united as they are from the same shell. In a sense, the security of marriage is the two halves combined inside a secure shell. In Gaelic France, walnuts are thrown at the couple, presumably as locally grown confetti. I assume they are thrown with care so as not to hurt the happy couple.

John Evelyn says in his book *Sylva* that "it would appear that in Germany no young farmer whatsoever is permitted to marry a wife till bringing proof that he hath planted a stated number of walnut trees as the law is inviolably observed to this day for the extraordinary benefit this tree affords to the inhabitance".

According to African American folklore walnut trees and leaves are used to put jinxes on people. They were also used to make people "fall out of love". Catherine Yronwode in her book *Hoodoo Herb Root Magic* describes a ritual where a tea of black walnut (husk and all) is made by boiling in three quarts of water: boiling it till the water evaporates down to one quart. You bathe in this water, renouncing ties to the former love, and then throw the water out at a crossroads or against a tree.

In literature the nut is sometimes used as a description of a person. In Arthur Ransome's book, *Peter Duck*, the *Swallow* crew of John, Nancy, Titty and Roger join the *Amazon* crew of Nancy and Peggy and Uncle Jim to set off on a voyage in the schooner *Wild Cat* half way around the world. They need one more crew member and find Peter Duck. He is introduced by Arthur Ransome as follows;

> "Peter Duck was sitting on a bollard on the North Quay of Lowestoft inner harbour smoking his pipe in the mid day sunshine and looking down at a little green two masted schooner that was tied up there while making ready for sea.
>
> He was an old sailor with a fringe of white beard round a face that was brown and wrinkled as a walnut. He had sailed in the clipper ships racing home with tea from China. He had sailed in the wool ships from Australia. He had been around the Horn again and again and knew it"

I wonder if Ransome meant the outer shell or the kernel inside.

The website *wow zone* in 2000 asks what tree did you fall from?

If you were born between April 21st and April 30th your tree is a walnut. You may be unrelenting, strange and full of contrasts. Often egotistical, aggressive, noble, broad horizoned, have unexpected reactions, spontaneous, unlimited ambition, no flexibility, difficult and uncommon partner, not always liked but often admired, ingenious strategist, very jealous passionate no compromise.

There are jokes about walnuts. Islamic humour relates the joke about the walnut and the pumpkin. One hot day, Mulla Nasruddin was taking it easy under the shade of a walnut tree. He noticed a huge pumpkin growing on a vine and the small walnuts growing on the tree above him. Why should a tiny walnut grow on such a mighty and majestic tree? As he thought about this, a walnut fell on his head without injuring him. He raised his hands and face to the heavens and said, "My god, forgive my questioning your ways! Where would I be now if pumpkins grew on trees?"

There is a city called Walnut in California. It covers about nine square miles and the name is derived from the Spanish *Ranchos los Nogales*. The translation of the word *nogales* is walnut.

Conclusion

The Walnut Tree

Once summer green
Then autumn gold
Bedecked this ancient tree –
Today silhouette laid bare
Mistletoed in winter frieze.

Brian Strand

The walnut tree is green, beautiful and abundant in summer and a magnificent shape in winter. It is one of the few trees that has both food value in its nuts and then, when its life's fruitfulness has come to an end, whose timber has a valuable after-life which can extend for many centuries. The walnut tree enriches so many aspects of life: landscape, art, furniture, sport, cooking, the list goes on. The unique smokey pattern in the wood is highly prized by cabinet makers and gunstock makers as well as the Jaguar car company. Its popularity never fades.

Walnuts have been grown in the United Kingdom since the Romans arrived, and there are many people mentioned in this book who have become enthusiastic growers and promoters of the walnut for different reasons. The long-term project at Lount to produce fine-quality walnut timber for dashboards in Jaguar cars, is an investment for the enjoyment of future generations; walnut furniture enhances some of Britain's finest houses; the health benefit of the nut in lowering cholesterol is scientifically proven.

I have spoken with many people who tell me they have walnut trees in their gardens. Some are lucky enough to have inherited well-established trees, but an ever-increasing number have planted theirs recently so are still waiting to be rewarded by future harvests.

My own walnut grove has given me a lot of pleasure. I take pride in the way the trees have grown and the quantity of nuts they produce. I like to keep the grove neat and tidy in the same way the French farmers do with their commercial groves. Perhaps these farmers have followed the good example of their neighbours' traditional vineyard husbandry?

 **Selective bibliography**

Adams, Max, *The Wisdom of Trees*, London, 2007

Bean, W.J., *Trees and Shrubs Hardy in the British Isles*, English 5th edition, London, 1925

Bowett, Adam, *English Furniture, 1660–1714: from Charles II to Queen Anne*, Woodbridge, Suffolk, 2002

Brickell, Christopher and Joyce, David, *Pruning and Training*, The Royal Horticultural Society, 2004

Brown, George E., revised by Tony Kirkham, *The Pruning of Trees, Shrubs and Conifers*, Cambridge and Portland, Oregon, USA, 1972

Campbell-Culver, Maggie, *A Passion for Trees: The Legacy of John Evelyn*, Bodelva, Cornwall, 2006

Campbell-Culver, Maggie, *The Origin of Plants*, London, 2001

Cooper, Jacob Alvin, *Walnut Growing in Oregon*, Portland, Oregon, 2010

Dallas, Donald, J*ames Purdey & Sons: Gun and Rifle Makers*, Shrewsbury, 2013

Davies, Brian, *The Gardeners Illustrated Encyclopaedia of Trees and Shrubs*, London, 1987

Desmond, Ray, *Sir Joseph Dalton Hooker*, Woodbridge, Suffolk, 2007

Dori, Jonathon, *Around the World in 80 Trees*, London, 2018

Hart, Cyril Raymond Charles, *British Trees in Colour*, London, 1973

Hemery, Gabriel and Simblet, Sarah, *The New Sylva: A Discourse of Forest and Orchard Trees for the Twenty-First Century*, London, 2014

James, N.D.G., *The Arborculturist's Companion*, Malden, USA, 1990

Johnson, Owen, *Champion Trees*, London, 2011

Johnson, Owen and More, David, *Collins Tree Guide*, London, 2004

Lovato, Kimberley, Schmalhorst, Laura and Lesko, Leo, *Walnut Wine & Truffle Groves: Culinary Adventures in the Dordogne*, Philadelphia, 2009

Mabey, Richard, *A Cabaret of Plants*, London, 2016

Macquoid, Percy, *A History of English Furniture: Volume 2*, New York, 1972

Milner, J. Edward, *The Tree Book. The indispensable guide to tree facts, crafts and lore*, London, 1992

Mitchell, Alan and Wilkinson, John, *The Trees of Britain and Northern Europe*, London, 1998

Mazet, Christian, *Je découvre, Tu découvres 'La Noix'*, Brive-la-Gaillard, 2010

Pakenham, Thomas, *The Company of Trees*, London, 2015

Phillips, Roger, *Trees in Britain, Europe and North America*, London, 1978

Potter, Jennifer, *Strange Blooms: The Curious Lives and Adventures of the John Tradescents*, London, 2007

Press, Bob and Hosking, David, *Trees of Britain & Europe (Photographic Field Guide)*, London, 1993

Prevost, John F., *The Walnut Tree*, Minnisota, USA, 1996

Schulz, Bernd, *Identification of Trees and Shrubs in Winter using Buds and Twigs*, Kew Gardens, London, 2018

Strang, Jeanne, *Goose Fat and Garlic: Country Recipes from South West France*, London, 2003

Wulf, Andrea, The *Brother Gardeners*, [?], 2009

Books with good walnut recipes (recommended by Jill Norman)

The Fruit and Nut Book
Helena Radeska, Sphere Books, 1984

Chez Panisse Fruit
Alice Waters, HarperCollins, 2002

Jane Grigson's Fruit Book
Michael Joseph/Penguin, 1982

Riverford Farm Cook Book
Guy Watson & Jane Baxter, 4th Estate, 2008

A New Book of Middle Eastern Food
Claudia Roden, Penguin, 1985

Tasting Georgia
Carla Capalba, Pallas Athene, 2017

Samarkand
Caroline Eden & Eleanor Ford
Kyle Books, 2016

Feast, food of the Islamic world
Anissa Helouo, Bloomsbury, 2018

Author's acknowledgements

Many people have helped me with my research. In particular I thank my family and friends for their advice, tales, stories and facts; Robin Balfour for introducing me to the Walnut Tree; Richard Purdey for introducing me to the Purdey Factory; Richard Scott for photographs; Phillip Wates, Head Gardener at Wimpole Hall; Tim Cornish, historian of Mayfield; Doctor Jo Crook; Bill Cleydert, veneerer; Dave Aidey, Veneer Manufacturing Centre; Mark McCarthy, Purdey; the staff at Holland and Holland; the staff of the Royal Armouries, Leeds; Tony Kennedy, gunsmith; Angus Ferguson, Demi-John; the staff at the Tomada Restaurant, St John's Wood; Karen Coke, curator at Chichester and expert on Lambert Bernard; Nick and George Newton, designers; Hugh Tempest-Radford, my publisher.

Picture credits

Cover: Getty Images
Chapter opener vignettes: iStock by Getty
    Images / KevinDyer
Page 28: Museumslandschaft Hessen Kassel,
    Wikimedia Commons
35, 36: Vatican Museum, Rome
37: © Florilegius / Science & Society Picture
    Library
57: Wikimedia Commons
58: Richard Scott
78: Hever Castle
82 (upper): Victoria & Albert Museum
    (lower): The National Trust
84: Milan Museum
85: Milan Museum

88, 89: The National Trust
91, 92, 94: VMC
97: Wikimedia / Prado
97 (lower): Brera Museum, Milan
99: Royal Armouries
100: (lower): Brera Museum, Milan
105: Holland & Holland
108: Purdey
109: Royal Armouries
111: Tony Kennedy
172: The Dean and Chapter, Chichester
    Cathedral
174: National Gallery, London
175: Bridgeman / Fitzwilliam Museum,
    University of Cambridge
176 (upper): Detroit Institute of Art
    (lower): Wikimedia Commons
177: Ashmolean Museum, University of Oxford
178: Wikimedia Commons

Other photographs by the author

Index